On the Shoulders of Giants

Other popular science books by Malcolm Lines available from
Institute of Physics Publishing

A Number for Your Thoughts
Facts and Speculations about Numbers from Euclid to the Latest
Computers

Think of a Number

On the Shoulders of Giants

Malcolm E Lines

Published in 1994 by
Taylor & Francis Group
270 Madison Avenue
New York, NY 10016

Published in Great Britain by
Taylor & Francis Group
2 Park Square
Milton Park, Abingdon
Oxon OX14 4RN

International Standard Book Number-10: 0-7503-0103-1 (Softcover)
International Standard Book Number-13: 978-0-7503-0103-9 (Softcover)
Library of Congress catalog number: 94-30118

Library of Congress Cataloging-in-Publication Data

Catalog record is available from the Library of Congress

informa
Taylor & Francis Group
is the Academic Division of Informa plc.

Visit the Taylor & Francis Web site at
http://www.taylorandfrancis.com

To Kathy

*A mathematician, like a painter or poet,
is a maker of patterns. If his patterns
are more permanent than theirs, it is
because they are made with ideas.*

G H Hardy

Contents

1

Introduction

Perhaps Oscar Hammerstein II, in conjunction with composer Richard Rodgers, said it most melodiously (albeit in a most unscientific setting) when he put the words 'nothing comes from nothing' into a ballad in the 1950s musical *The Sound of Music*. Or was it the irrepressible Harvard mathematics professor and songster Tom Lehrer who, in that same decade, expressed it in the more cynical form 'plagiarize, plagiarize, let no-one else's work evade your eyes' in his irreverent jingle concerning the supposed advice of the (actually much respected) nineteenth century Russian mathematician Nikolai Ivanovich Lobachevsky on how best to get ahead in research endeavors? In any event, the sentiment that none of us creates anything truly worthwhile entirely out of a vacuum is not one restricted to musical plays or satire. In a much more serious vein, and in better accord with the theme of this book, this has appeared in many and diverse forms at least as far back as Roman times, and probably much farther.

The most famous quote along these lines is almost certainly the one usually attributed to Sir Isaac Newton who, in a rather uncharacteristically modest fashion, spelled it out in a letter to his distinguished scientific compatriot Robert Hooke in 1675 in the form 'If I have seen farther [than others] it is by standing upon the shoulders of giants'. However, this oft-quoted remark is not a Newtonian original. One can find essentially the same statement in the historical epic *Bellum Civile* (the Civil War) by the first century Roman poet and republican patriot Marcus Lucanus in words that literally translate as 'pigmies placed on the shoulders of giants see more than the giants themselves'. The sentiment is the same, but one can see why Newton chose to paraphase the original.

Although it should come as no surprise to anyone that research in any field of endeavor conventionally builds upon the existing 'pool of knowledge' already established for that field— something we like to refer to as 'pushing back the frontiers'—we shall attempt to establish in the following chapters the existence in the physical sciences of something that often goes, rather mysteriously some might think, significantly beyond this assertion. In these chapters we attempt to demonstrate the manner in which many of the most significant advances of twentieth century physics have found the mathematics which is ideally suited (and sometimes essential) for their theoretical understanding quietly lying-in-wait (almost in the nature of a solution looking for a problem) in the mathematics literature of earlier generations. In other words, the giants whose shoulders are most desperately required by research scientists often turn out to be those of long-ago-expired mathematicians. Mathematicians, on the other hand, would claim that they are rarely, if ever, to be found standing on the shoulders of scientists.

To many it may not seem surprising that mathematics has played an essential role in science, and particularly physics— after all, physics is a quantitative science. Nevertheless, mathematics and physics are by their nature very different disciplines indeed. In fact, mathematics possesses a methodology that is unique among all the sciences, since it is the only science in which deductive logic alone determines truth. It follows that mathematical theorems which were established thousands of years ago are every bit as true today as they were on the day that they were first proved. In many cases they are still taught as part of the essential curriculum in today's High School mathematics courses. Not for the mathematician are the messy uncertainties concerned with instrumental errors, statistical accuracy of data, or 'tentative' assumptions that so often go along with potential advances in other sciences. For this reason, mathematicians tend to believe that the quality of their truths is higher than that of other sciences and are justly proud of their standards of rigor.

As a humorous illustration of this rigor, I am particularly fond of the story concerning three academics supposedly traveling by train from London to Edinburgh to attend a conference. Upon crossing the Scottish border, they casually gazed out of the train window to see two lone black sheep in a field. The first academic, an astronomer by discipline, expressed immediate interest. 'Just look at that,' he said, 'the sheep in Scotland are black.' The second academic, being a physicist, was jarred by this unwarranted generalization. 'That is nonsense,' he responded. 'All that the evidence before us demonstrates is the fact that at least two of the sheep in Scotland are black!' After a moment or two of silence, during which time the physicist basked in the glory of his superior powers of reasoning, he found himself equally rebuked by the third member of the group, a mathematician. 'Your conclusion is equally unfounded,' the mathematician retorted. 'The correct interpretation of the observation is that at least two of the sheep in Scotland each have at least one black side!' Although the story is perhaps a little unkind to astronomers, it does focus on an important point. The physicist, never having seen or heard of the existence of a sheep anywhere on earth with one black side and another side of a different hue, was happy to accept the 'fact' that none exists. To the mathematician, in the absence of any proof that no such animal could possibly exist, such an assumption was completely unwarranted.

Other somewhat more serious demonstrations of this difference in attitude between physicists and mathematicians are not difficult to highlight—examples which again focus on the uniqueness of mathematical rigor. For instance, it is often said that of all scientists, the mathematician is the only one who does not know how to pack spheres together to form the densest possible arrangement in three-dimensional space. All physicists, for example, 'know' that the answer is just the ordered stacking that you often see for oranges in supermarket displays or for cannonballs at War Memorials. It is called 'ordered close packing' and fills up very nearly 75% of all space if continued *ad infinitum*. The mathematician, looking at the same problem, is quite happy to accept that this is, indeed, the densest ordered (that is, repeating or periodic) arrangement, since a rigorous proof of the fact exists. However, as yet, no mathematical proof has been given that some non-periodic arrangement could not do better. In truth, the physicist also initially had doubts, and so he experimented—first by shaking balls into containers and later, after the advent of the computer, by creating non-periodic algorithms (or sets of rules) to 'build' such packings by computer simulation. But try as he would, over years and years of effort, he found that no randomlike packing could ever be assembled with a packing density which came even close to the 75% of space filling achieved by the periodic orange and cannonball arrangements. In fact, after decades of trying, only about 65% of space could be filled by even the cleverest computer-prepared non-periodic packings, a value not significantly greater than the one achievable by simply shaking balls in any large container. Thus, for the physicist the question is settled 'beyond a reasonable doubt'. But for the mathematician, even an unreasonable doubt is not sufficient. In the absence of a proof that satisfies the full rigor of mathematical methodology, the question of the most efficient packing of identical spheres in three dimensions remains unanswered.

In a similar vein, the late physics Nobel Laureate Richard Feynman once presented a physicist's proof of Fermat's Last Theorem. This most infamous of all (unsolved) mathematical conjectures states simply that the equation $x^n + y^n = z^n$ cannot be satisfied by any positive integer values for x, y, z and n if the exponent n is larger than 2. I place the unsolved adjective in parentheses since, after some 350 years, a rigorous mathematical proof which may possibly withstand the scrutiny of the refereeing panel has been presented during the writing of this book. It was offered by Andrew Wiles, of Princeton University, US, and announced to a stunned audience in a lecture hall at the Isaac Newton Institute in Cambridge, England. All 'proofs' until now have failed to hold up, and Wiles' effort is at present undergoing a 'fine toothcomb' examination by experts. However, initial comments—such as 'the logic is utterly compelling', 'it simply has the ring of truth' and 'even if there's a small mistake, it's likely to be fixable'—seem most encouraging so that, by the time you read this, the question may finally have been settled. But this 'breaking' news story has forced me to digress from my main point, the 'physicists' proof'.

Even in the days of Feynman's comment, the conjecture had been verified by computer up to extremely large integers, so that Feynman decided to make use of the random occurrence of numerals in still larger numbers to calculate the statistical probability that the Fermat equation might be satisfied. In this fashion he concluded that, even when integers of all possible digit lengths (beyond those already tested) were included, the probability that any particular set would satisfy the Fermat equation was much smaller than a physicist would be more than happy to accept as conclusive for any physical experiment. Thus he asserted (in characteristically impish fashion) that physicists should be just as happy to accept the validity of his 'proof' of Fermat's Last Theorem as they are to accept any experimental 'conclusion' resulting from their discipline.

Mathematicians would obviously be aghast at such an approach to the problem. Assertions concerning the randomness of digits in large numbers are dubious in the extreme. Problems similar to Fermat's, which *have* been unquestionably solved and are known to have an *infinite* number of possible solutions, are

A young Albert Einstein at his desk in the Patent Office in Berne ca 1905. Courtesy of The Hebrew University of Jerusalem, Einstein Archives. Reproduced by permission of AIP Emilio Segrè Visual Archives.

also known to pass the Feynman test for insolubility with precision comparable to the finding for the Fermat equation itself. Evidently the sciences of physics and mathematics possess fundamental differences in their approaches to the 'search for truth'. In a very real sense, mathematics is divorced from experiment, while physics appeals to experiment as the sole arbiter of truth. This difference is illustrated by an oft-told (though possibly apocryphal) meeting between the youthful Albert Einstein and an aging Jules Henri Poincaré, the latter of whom was one of the foremost mathematicians of the nineteenth century. Einstein supposedly said to Poincaré, 'I considered taking up mathematics but decided against it because of its lack of connection with the real world and the impossibility of telling what is important.' To

Jules Henri Poincaré, 1854–1912. Reproduced by permission of Mary Evans Picture Library.

which Poincaré replied, 'In my youth I was seriously tempted to become a physicist, but I decided against it because in physics it is impossible to tell what is true.' Clearly, if the two subjects are to reach compatibility then some trade-off has to be made between the high standards of truth and the requirements of applicability.

As a very simple demonstration of this trade-off let me discuss for a moment the subject of the science of crystallography. By various experimental techniques involving the scattering of x-rays or neutrons from crystals, it is nowadays a common procedure to (as physicists put it) 'determine' the 'structure' of a particular crystal. By this they mean identify the shape and size of the smallest atomic unit which repeats itself endlessly (or at least of order 10^{23} times) in a regular pattern to build up the macroscopic sample. As a simple example we can imagine 10^{23} cubes packed together with atoms at the cube corners and, perhaps also, the cube centers or cube face centers. Many of the most common metals in the world have structures of this kind, and here we would appear to have a crystallographic truth that is perfectly representable by mathematics. But we have deceived ourselves. The laws of thermal physics dictate that no such perfect 'mathematical' crystal can possibly exist at any non-zero temperature—and the absolute zero of temperature can never be attained. In all actual crystals there must, in addition to atomic vibrations about their ideal sites, be defects—that is, some atoms absent from the sites where they should be, and others in displaced sites where they ideally should not be. Clearly the real crystal is not exactly represented by the mathematical lattice (and can never be, even in ideal physical circumstances). And yet, since the vast majority of atoms are indeed where they 'should be', the mathematical description is of great value. In other words, a valuable trade-off between mathematics and physics has been made. In like manner there is much more cooperation than confrontation between the physical sciences and mathematics than their different visions of truth might suggest.

In truth, if one factor has remained constant throughout the twists and turns of the history of progress in the physical sciences, it is the decisive role played by mathematics. It is the story of this role in many of its guises that the chapters of this book will attempt to portray. In many instances, a case can be made for the statement that it is the mathematics that drives the physics. Indeed, the power of existing mathematics to mirror the behavior of the universe has ever been a source of wonder to physicists, and has even attracted the concern of philosophers. The usual procedure is for the physicist to carry out a translation of the physical reality into the abstract language of mathematics by locating the most 'relevant' mathematics available for the purpose. Surprisingly often, this turns out to be some previously unapplied corner of purest mathematical abstract enterprise. Application of this existing mathematical framework to its new context then not only assists enormously in clarifying the thinking surrounding the physical phenomenon, but also frequently points the way to fruitful directions in which the physical research in that area can be extended.

Many physicists over the years have expressed their amazement at the potency of this 'marriage'. The nineteenth century German physicist Heinrich Hertz, after using differential equations in pursuit of an understanding of radio waves, wrote that 'One cannot escape the feeling that these mathematical formulae have an independent existence and intelligence of their own; that they are wiser than we are, wiser even than their discoverers, and that somehow we get more out of them than was originally put in.' Later, in the twentieth century, the same feeling of wonder was expressed by Eugene Wigner, the Hungarian-

born American physics Nobel Laureate, when he contemplated
the successes of various mathematical concepts in the interpretation of quantum mechanics, the physics of the weird and
wonderful world of the atomically small. In reference to this
mystical and almost supernatural association between the newly
discovered physical world and the previously unapplied abstract
mathematical formulations he wrote 'We are in a position similar
to that of a man who was provided with a bunch of keys and who,
having opened several doors in succession, always hit on the
right key with the first or second trial. He became skeptical
concerning the uniqueness of the coordination between keys and
doors.' Not to suggest, of course, that all physicists (whether
experimentalists or theorists) have been quick to recognize the
mathematical 'jewels' offered to them. The British physicist and
astronomer Sir James Jeans, while Professor of Applied Mathematics at Princeton in 1910 (and before his elevation to knight-
hood in 1928) is purported to have advised the curriculum
committee 'We may as well cut out group theory. That is a subject
which will never be of any use in physics.' He could not have
been more mistaken (as we shall see later in the book) and this
same group theory, a product of the purest mathematical
thought, is now not only the essential language of the physics of
crystalline solids, but also completely dominates the thinking of
all who are trying to understand the fundamental particles and
forces of nature at the subatomic level.

So how do we explain this uncanny ability of formal and
abstract mathematics to come to the aid of physical scientists in
their efforts to understand the world around them? One recent
suggestion by a Harvard mathematician goes something like this:
'Mathematics is the science of patterns; patterns in number, in
space, and in imagination. Theories in mathematics then pursue
the relations among these patterns, binding one to the other to
yield structures. Applications of mathematics to the physical
sciences use these patterns to "explain" natural phenomena that
fit the patterns. In recent years, particularly with the advent of
computers (which are to computational mathematics what
microscopes are to science) the portfolio of these patterns has
been increased immensely. Could it therefore be that the reason
why mathematics has the uncanny ability to provide the right
patterns to explain natural phenomena is simply that the number
of patterns now recognized by mathematicians is fast approaching all the patterns that there are?' Although persuasive, perhaps,
as an argument as to why mathematics may continue to fulfill this

role in the future, I personally am not impressed by it as reason for its uncanny success in the pre-computer world of the distant, and not so distant, past. At present I am content to let it remain an enigma, and proceed to relate the individual stories that constitute the evidence.

2

From Aristotle to the Structure of Glass

The question of how to fill space, without leaving any gaps, by fitting together identically shaped building blocks is one of the most ancient of all problems about space. In one sense the answer is easy (the bricklayer, for example, has obviously discovered one solution) but in another it is extremely difficult. Which shapes of block are able to fill space in this manner and which are not? In its more general form, perhaps somewhat surprisingly, the question remains unanswered to this day, computer graphics notwithstanding. But interest in pursuit of solutions has produced an enormous amount of effort over the centuries which has, as we shall see, finally left the realm of the academic and become firmly enmeshed in the study of glass structure; glasses being those solids that seem to have trouble in finding a way to pack space without becoming scrambled and distorting their atomic 'building blocks'.

Questions concerning the filling of space with identical blocks appear to have first arisen near the beginning of the fourth century BC in connection with a theory of matter advanced by the Greek philosopher Plato (circa 427–347 BC). This theory contained an hypothesis that all matter is composed of an assembly of space-filling polyhedral units, polyhedra being geometric solids with plane polygons (e.g. triangles, quadrilaterals, etc) as faces. In this context Plato placed special emphasis on what he called the five regular solids, since they (in his eyes) were the block shapes which most closely represented perfect symmetry in solid geometry. By definition, a regular solid is a polyhedron with all its faces identical in size and shape, and with all the angles between adjacent faces also equal. The most obvious example is a cube, for which all the faces are equal-sided squares

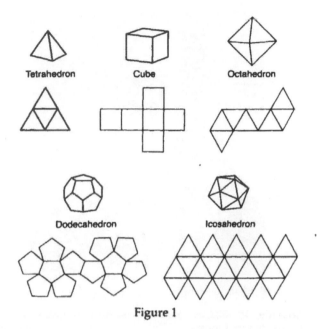

Figure 1

and all adjacent face angles are ninety degrees. The definition of regular solids, and the proof that there are only five of them in three-dimensional space, was one of the great mathematical accomplishments of the Ancient Greeks. These five solids, with respectively, 4, 6, 8, 12, and 20 faces, are shown in figure 1, where we also indicate the paper construction patterns from which they may easily be assembled.

Although, as far as we know, Plato was the first to qualitatively envisage the packing of space with regular solids (with the idea that some, or all, of these very special units might be the ultimate 'atoms' from which all matter is composed), it was Aristotle (384–322 BC) who first tried to examine the details of the associated packing problem. Aristotle, who was one of Plato's more notable students, argued correctly that his tutor's idea could not be fully compatible with reality since most of these fundamental solid shapes could not be packed together to fill space without leaving gaps. More precisely he stated that, of the five regular solids, only the six-sided cube (for which the associated packing is easy to envisage) and the four-sided tetrahedron (for which the packing is not so obvious) could separately pack together to fill space completely. The others would leave gaps—and gaps (according to Plato's theory) could not exist within matter.

Aristotle, 384–322 BC. Reproduced by permission of Mary
Evans Picture Library.

Now, as it happens, this more detailed statement from Aristotle
is incorrect. In fact, of the regular solids, only the cube will exactly
pack together to completely fill space in the fashion required.
Regular tetrahedra (figure 1) can *almost* do the job, but not quite.
They always leave small spaces between the pieces no matter
how hard you try to persuade them to fit together. Why, you may
ask, didn't Aristotle construct some regular tetrahedral blocks out
of actual material and try it for himself? We don't know for sure,
but this probably gives us some indication of the low opinion that
the Greeks had of the relevance of messy approximate experimentation in the context of mathematical enterprise. Nevertheless, for whatever reason, Aristotle's mistake was not noticed at
the time and such was his stature that, although to our knowledge he provided no evidence for his claim, if any later scholars
had doubts about the assertion they never recorded it. Unbelievably, this misconception that regular tetrahedra can pack
together to fill space persisted for some *eighteen centuries*, and
even after its resolution, this Aristotelian error continued to
appear in various guises for many years. This is truly amazing,
implying that experimentation played virtually no role in probing

the question for all this time—for although we realize that a *proof* of the truth of this or any other geometric assertion can only be accomplished mathematically, a little practical work would quickly have raised great doubts!

The essential problem is that 'visualizing' the three-dimensional packing of regular tetrahedra is quite difficult. The equivalent problem of packing together regular polygons in two dimensions is much easier, in spite of the fact that there are an infinite number of them (n-gons, if you like, starting with the equilateral triangle for which $n = 3$, the square with $n = 4$, the regular pentagon with $n = 5$, and so on for $n = 6, 7, 8, \ldots$) compared with only five regular solids. Let us think of a way in which we might attack the problem of completely filling two-dimensional space (which we shall call 'the plane') with identical regular polygons. Quite obviously, one essential condition is that it must, at the very least, be possible to fit them together precisely around a point at which they touch. Which n-gons will exactly fit together at a point? To answer this question we need to find the interior angle (that is, the angle between two adjacent sides on the inside of an n-gon). If we call this the angle A, then the condition for perfect packing around a point is simply that the equation $mA = 360$ (degrees) has a solution for which m is a whole number. Quite obviously one solution can be found for the square, with $A = 90$ degrees and $m = 4$. In other words, four squares can be perfectly fitted together to share a corner.

But what of the general problem for identical n-gons? What is the identity of all the n-gons that can tile a plane in the manner prescribed? In order to attack this problem we first need to find the interior angle A for the general n-gon. To do this it is easiest to imagine taking a match stick and moving it steadily around the edge of the n-gon until we get back to the starting point. In its travels, the match stick has obviously rotated through one complete revolution (or 360 degrees) and it has performed this rotation in n equal-angle steps. It follows that, at each corner of the n-gon, the match stick has tipped by $360/n$ degrees. Consequently, the inside (or interior) angle at each 'corner' is just 180 degrees minus this amount. In symbols this becomes

$$A = 180 - (360/n).$$

The condition for perfect packing around a point is now

$$mA = m[180 - (360/n)] = 360$$

which, after a little algebraic tidying up, can be rewritten in the form

$$\frac{1}{m} + \frac{1}{n} = \frac{1}{2}.$$

Our 'solutions' are the whole numbers m and n which can satisfy this equation. Trying in sequence the n-gons of increasing 'size' $n = 3, 4, 5, 6$, etc, we quickly find the solutions $n = 3$ ($m = 6$), $n = 4$ ($m = 4$) and $n = 6$ ($m = 3$). Since m must get smaller as n increases (a fact that is evident from the last equation) the next possible solution, if it exists, must be for $m = 2$. But this requires an n-gon with an infinite number of sides and an interior angle of 180 degrees, and clearly represents an unphysical limit. It follows that there are only three regular polygons which are candidates for 'tiling' the plane completely with their own kind. These are: (i) the equilateral triangle, with $A = 60$ degrees; (ii) the square, with $A = 90$ degrees; and (iii) the regular hexagon, with $A = 120$ degrees. We have so far demonstrated only that these three can fit together precisely about a point. That they all do, in fact, completely fill or tile the plane when the pattern is continued is easily demonstrated pictorially, as shown in figure 2.

In three dimensions we can ask corresponding questions. For the regular tetrahedra that were the focus of Aristotle's troubles the first question now becomes 'How many of these tetrahedra can be packed together to share a common corner, and is this packing complete in the sense of leaving no gaps?' I think that you will agree that visualization of this situation in the mind's eye is quite difficult (unlike the analogous problem for cubes, where eight cubes fit together perfectly to share a common corner). The

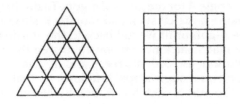

Figure 2

first quantitative statement known concerning this particular numerical problem was made by Simplicius, a commentator and scholar who lived in the sixth century. He asserted that the number is 12, and implied that the packing is complete, although he gave no proof. That something was wrong with this conclusion was pointed out by the Oxford academic Thomas Bradwardine (1295–1349) who noted that if 12 regular tetrahedra did indeed exactly fill the space about a point in three dimensions, then altogether they would define a sixth regular solid (one with 12 equilateral triangular faces) to add to the five Platonic originals shown in figure 1. Since the proof of the existence of only five regular solids was already firmly established, the statement of Simplicius had to be wrong. Bradwardine, himself, was more inclined toward the answer 20, since 20 tetrahedra packed together at a point would make up an icosahedron (figure 1). On the other hand he had doubts as to whether these particular component tetrahedra would be regular (they are not) or whether the resulting packing arrangement could be repeated to fill all space (it cannot) since, as Bradwardine was well aware, filling space about a point is, in itself, no guarantee that the packing can be continued to fill all space.

What was required was a return to the geometry of Euclid (see chapter 3) which had been widely neglected after the demise of the early Greek civilization. Only in the fifteenth century, when Euclid once again began to be studied, did the confusion surrounding the packing of regular tetrahedra begin to move towards a resolution. Again, we can only express amazement that neither Bradwardine, nor to our knowledge anyone else during this lengthy period of controversy, made the effort to model regular tetrahedra, after the fashion of figure 1, in order to check out the situation experimentally. Although such model-fitting, as we have stressed above, cannot prove anything in a rigorous fashion, it can be very suggestive—and certainly rules out any notion of a packing number of 20. In fact, if you try it, you will quickly find out that 12 regular tetrahedra can certainly share a common corner—but there are gaps, and these gaps do not seem to be quite large enough to squeeze in a thirteenth tetrahedron. Thus, in spite of its crudeness, this little insertion of messy 'experimentation' into the higher world of mathematics does seem to provide very persuasive 'evidence' that regular tetrahedra can never pack together to fill space— in spite of Aristotle's assertion to the contrary.

The first mathematical demonstration of this truth is thought to have been given by the German mathematician and astronomer

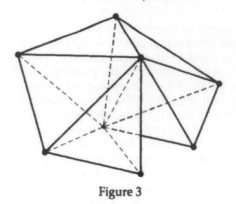

Figure 3

Johannes Regiomontanus (1436–1476), the author of an important work on spherical geometry (that is, geometry on the surface of a sphere), although the proof itself has not survived in its original form. In spite of this, by a return to the rigor of Euclid, it is quite clear that the question had been unequivocally settled by the mid-1600s at which time the Polish mathematician Jan Brozek (1585–1652) published a book in which the problem of filling space with regular solids was discussed correctly and in detail. And of the five such solids, only the cube could do the job.

Now once you know the secret, it turns out that the proof that regular tetrahedra cannot possibly fill space is not at all difficult. In fact, it is quite simple enough to set out here. The secret is to focus not on the packing about a common point (or 'vertex'), but rather about a common edge, as shown in figure 3. If we can calculate the angle x between adjacent faces of the regular tetrahedron (the dihedral angle; see figure 4(a)), then these tetrahedra cannot possibly fill space unless a whole number of them can exactly pack together to share a common edge; that is, unless the equation $mx = 360$ degrees has a whole number solution for m.

It just so happens that the dihedral angle x is extremely simple to calculate. Consider the regular tetrahedron ABCD shown in figure 4(a). If we bisect the two equilateral triangular faces ABC and BCD by drawing the lines AE and DE to meet BC at right angles, then clearly the dihedral angle x is the angle DEA. Suppose that the tetrahedron ABCD has side lengths equal to 1. We already know, therefore, that BE = EC = $\frac{1}{2}$. Also, since triangles AEC and DEC both contain right angles (figure 4(a)) we can use the Pythagorean theorem to calculate the lengths

$$AE = DE = \sqrt{(1^2 - \tfrac{1}{2}^2)} = \sqrt{3}/2.$$

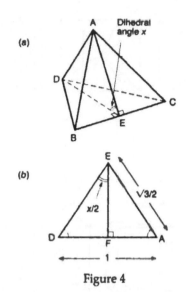

Figure 4

The dihedral angle x is therefore the apex angle of an isosceles triangle (DEA) with side lengths 1, $\sqrt{3}/2$, $\sqrt{3}/2$. By dropping a perpendicular EF from the apex E to the base DA (see figure 4(*b*)), we immediately see (since DF = FA = $\frac{1}{2}$) that one half of the dihedral angle ($x/2$) is the angle whose sine is FD/DE = $\sqrt{(1/3)}$ = 0.577 350. This angle, according to my calculator, is 35.264 389 68 degrees, so that the full dihedral angle is twice this or x = 70.528 779 37 . . . degrees. Now, quite clearly, when this is substituted into the equation mx = 360 degrees, it does not give a whole number solution for m. It actually gives a number that starts m = 5.104 299 31 . . . . It follows that five regular tetrahedra can share a common edge but that, just as depicted in figure 3, a small gap or wedge, of a little more than 7 degrees, remains unfilled. We conclude that regular tetrahedra cannot possibly pack together to completely fill space.

There is no implication in any of this, of course, that cubes are the only solids which, by themselves via constant repetition, can completely fill space. If we relax the limitation to 'regular' solids then, as the bricklayer is well aware, more generally shaped rectangular solids will also do the trick. Nor, as we shall see, is any restriction to solids with rectangular corners or with six faces essential. A serious interest in the more general space-packing problem came with the dawn of the science of crystallography. The latter is most often traced back to the French abbot and

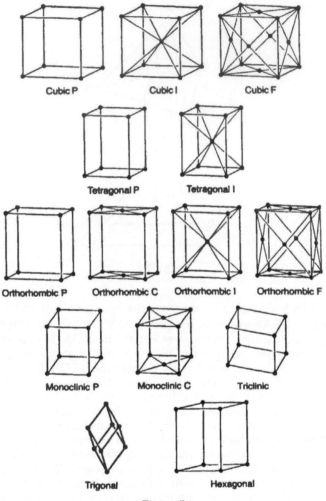

Cubic P Cubic I Cubic F

Tetragonal P Tetragonal I

Orthorhombic P Orthorhombic C Orthorhombic I Orthorhombic F

Monoclinic P Monoclinic C Triclinic

Trigonal Hexagonal

Figure 5

mineralogist René Haüy (1743–1822) who pioneered the idea that the smallest repeating building blocks of crystals could be thought of as 'crystal molecules'. These are not necessarily chemical molecules as such, but are the smallest atomic groupings from which the macroscopic crystal can (at least in principle) be built up by straightforward assembly. If each of these molecular blocks (whatever they are) is replaced by a point at its

Auguste Bravais, 1811–1863. Courtesy of the Photographic Library of the Laboratoire de Minéralogie-Cristallographie, Université Pierre et Marie Curie.

geometric 'center of gravity', then a spatial pattern of points is created which is known as a space lattice. A lattice of this sort is very special since, by its definition, each point has an identical environment (the array of points being imagined to extend to infinity). The mathematicians refer to such a set of points as being 'translationally invariant' since by moving (or translating) the entire rigid lattice without rotation until one point coincides with another, then the lattice appears to the eye to be completely unchanged.

Although speculations concerning the relationship between the shapes of crystals and their possible internal structure had been discussed as far back as Kepler (1571–1630), studies during the eighteenth century focused primarily on the external forms alone, which were after all observable in nature. After Haüy, however, attention was again directed to the unobservable inner structure. The establishment of the existence of translationally invariant lattices (as defined above), and the fact that there are only 14 fundamentally different types (see figure 5), was the outcome of the work of several mathematicians in the first half of

the nineteenth century. Pre-eminent among these was the
French botanist and physics professor Auguste Bravais (1811–
1863) who, in an exhaustive study of the properties of lattices,
provided the final rigorous demonstration of the sufficiency and
completeness of the 14 types in 1848. They are now known as
'Bravais lattices' in his honor.

In addition to the space-filling molecular blocks (or 'unit cells')
which can be defined by joining any lattice point to its three non-
coplanar nearest neighbors (as generators), other more fascinating space-filling 'solids' can also readily be found. For example,
one can define a region of space which is closer to a given lattice
point than any other. These elementary solids are called 'paralle-
lohedra' and are clearly, by their very definition, related to each
other by translation in a manner which fills space completely
without gaps. They can be shown to be of five basic types (figure
6); a cube, an hexagonal prism, a rhombic dodecahedron (with 12
faces, each a rhombus), a dodecahedron with eight rhombic and
four hexagonal faces, and finally a truncated octahedron (that is,
an octahedron with each of its eight corners lopped off). Clearly
therefore, many weird and wonderful solid shapes can be
unearthed that will pack together to fill space without leaving any
gaps. The particular ones cited above (now referred to in the solid
state physics literature as 'Wigner–Seitz' cells), together with
their parent Bravais lattices, today provide the mathematical and
conceptual basis for many sophisticated theories of the electronic
properties of solids. They were also invaluable, in the early
twentieth century, for help in defining real crystal structures,

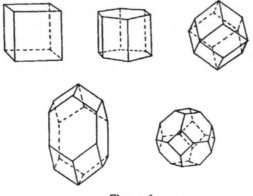

Figure 6

when the discovery of x-ray diffraction (by the German physicist Max von Laue in 1911) and its adaptation two years later for determining the atomic structures of crystals (by the British father and son team of Henry and Lawrence Bragg) finally opened up the realm of the atom to direct experimental measurement.

With all these shapes existing that can pack together perfectly to fill space, the reader may now be wondering why we spent so much time focusing on Aristotle's regular tetrahedron which cannot. The purist might reason that, at least in one sense, the tetrahedron is the simplest of all solids because it possesses the smallest number of faces, edges, and corners of all polyhedra. And it happens that some tetrahedra can be found (albeit not regular ones) which can fill space without leaving gaps. One such example can be obtained by partitioning the cube into six identical pyramids and then partitioning these pyramids further, each into four equal tetrahedra (see figure 7 for an illustration). But our reason for paying so much attention to the regular tetrahedron is the fact that four equal spheres, when packed together as closely as possible (that is, with each one touching the other three), have their centers at the corners of a regular tetrahedron. The problem of packing space with regular tetrahedra is then exactly analogous to the problem of packing space with equal-sized spheres

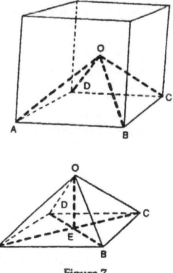

Figure 7

mentioned in the introduction. In particular, Aristotle's statement that regular tetrahedra completely fill space is equivalent to the statement that equal-sized spheres can be packed together in space in a 'perfectly dense' fashion. By this last statement we mean that, starting from a tetrahedral 'seed' of four equal spheres, space can be filled by adding other spheres sequentially in a manner such that each sphere added contacts three spheres to form a tetrahedron. The density of this packing, measured as the ratio of the space occupied by spheres to all space (and sometimes also referred to as the 'packing fraction') is about 0.7796. But the fact that Aristotle was wrong means that no such 'ideal' arrangement exists and the problem of packing equal spheres together in three dimensions to form the densest possible array now becomes one of immense complexity — so complex that to this day nobody knows how to do it, or what value the packing fraction f would then actually have.

From a practical point of view an approximate solution can be sought by simply shaking up a large number of equal-sized balls in a container (perhaps table tennis balls or golf balls in a bucket would do for a start). If you try this you will find the number of balls that can be accommodated (by shaking) in a large container is remarkably constant and gives a packing fraction of about $f = 0.64$. However, as noted in chapter 1, this is not the largest packing fraction possible for spheres by any means. Just as the bricklayer, who knows how to pack bricks together to completely fill space ($f = 1$), is likely to fail in this endeavor if he merely throws them randomly into a container, so the sphere-packer can do better than this randomly acquired $f = 0.64$ if he assembles the balls in an orderly fashion. The most densely ordered arrangement of spheres can be built up by first arranging three spheres on a flat surface, each touching the other two. Starting from this 'seed cluster' we continue the packing first by adding spheres on the surface such that each new one touches two of those already in place. In this manner we finally obtain an (in principle) infinite layer of spheres in which each touches six others. If we call this layer A, then we can build up a second layer on top of the first by placing spheres in the indentations or depressions left at the center of each triangle of spheres in the first layer. However, of the six indentations that surround a particular sphere in layer A, only three can be filled in the adjacent layer. There are therefore two separate ways of building up the second layer (which we refer to as layers B and C respectively) each differing from the other by a rigid horizontal translation. More layers can now be built on top of the first two in an analogous fashion, each in the indentations of the previous layer, until a full (infinite) three-dimensional arrangement is completed.

We find that there are only three different 'horizontal' layer positions, namely layers of type A, B and C. Nevertheless, this enables us to build up an infinite number of different stacking arrangements, all looking rather alike to the untrained eye. All have the same packing fraction $f = \pi/\sqrt{18}$ or, in decimals, a number beginning 0.740 48. Each of these packings is said to be 'ordered' although only two of them, the cubic close packing (with layer arrangement ABCABCABC ...) and the hexagonal close packing (with layer arrangement ABABAB ...) are truly translationally invariant in three dimensions. We see that the careful ordered positioning of balls has produced a handsome improvement in packing fraction over the 0.64 achieved by mere shaking.

The fact that the dense ordered packings of spheres described above are the densest possible ordered arrangements that can be achieved was proved by the German mathematician Karl Friedrich Gauss (1777–1855) in 1831. However, as mentioned in chapter 1, no-one to this day has managed to establish that some very cleverly assembled disorderly arrangement could not do better. The only rigorous result available for disordered arrangements is the rather weak one that f must be smaller than that corresponding to the unattainable Aristotelian tetrahedral packing limit of 0.7796 .... The fact that nobody has ever yet managed to build (either by physical model or by computer generation) any non-repeating pattern of spheres with a packing fraction anywhere near the ordered packing value (or, indeed, significantly larger than the $f = 0.64$ attainable by random shaking) strongly suggests that none exists. But this does not constitute a proof of any kind. Therefore, no matter how unlikely the possibility may appear, from a rigorous mathematical standpoint the question of the densest possible packing in three dimensions remains unanswered.

And what has all this to do with modern science? Well, there is a well known state of matter which does, in a sense, much more closely resemble balls thrown into a bucket than balls stacked carefully in an orderly fashion. It is the 'glassy' state. The analogy is not a particularly good one, as we shall see, for the most common glasses (such as window glass, for example) but it is not so unrealistic for certain kinds of metallic glass. However, let us not get too far ahead of ourselves. Before we probe the microscopic details of the glassy state and its relationship to the random packing of hard spheres, we need first to answer some rather basic questions. What, for example, is a glass?

Most people, I suppose, have some concept of what we mean by a solid and by a liquid, and recognize that they represent quite different states of matter. A solid is 'rigid' and therefore tends to stay where you put it. Liquids, on the other hand, flow; and, on unfortunate occasions, may stray significantly from where you thought you put them! The fundamental difference, at the microscopic level, revolves around the fact that atoms in liquids are able to move about without undue difficulty throughout the entire macroscopic volume, while atoms in solids can usually move only vibrationally in small-amplitude oscillations about fixed positions. Solids, therefore, can resist efforts to change their shape when subjected to external forces designed for that purpose (called 'shear' forces) while liquids cannot. This ability to resist change of shape (or more precisely flow) on application of shear forces is measured by a physical quantity called viscosity. In this language, liquids have low viscosities (with some, like water, having lower values than others, like maple syrup) while solids have viscosities that are close to infinite.

All liquids eventually transform into solids when the temperature is lowered far enough. High temperatures (or large thermal energies as we should perhaps more properly say) agitate the atoms and disorder them as much as possible, thereby generally favoring the liquid state over the solid one. However, attractive forces of interaction do exist between the atoms in the liquid and these, given low enough temperatures and sufficient time, will always finally arrange the atoms of a particular composition in that unique and essentially static pattern for which the sum of all these interactions is strongest (or, more precisely, for which the total energy resulting from these interactions is minimum). This particular atomic arrangement is almost always ordered (that is, translationally invariant) although there is no mathematical proof that this must necessarily be the case. Indeed, the recent discovery of what are called 'quasicrystals' (the details of which we shall turn to in chapter 6) strongly suggests that in some cases this may not be. In spite of this caveat, we may say that the stable state of almost all materials at low enough temperature is the ordered solid state known as a crystal.

At any particular temperature, therefore, there is a 'battle' between thermal energies (which favor disorder) and interatomic forces (which favor order) the outcome of which determines the actual state of matter at that temperature. When liquids are cooled, they therefore inevitably reach a temperature for which the thermal energies are no longer large enough to win the battle. At this temperature, and usually within a tiny fraction of a second, the system changes locally from liquid to crystalline solid. In common language it freezes, and the resulting solid is a crystal in which the atoms are all neatly arranged in a very specific repetitive fashion (all, that is, except for a few renegade atoms as discussed in chapter 1). The reason why an entire macroscopic sample (such as a block of ice in the refrigerator) does not change completely from liquid to solid in a tiny fraction of a second involves the time required for heat to flow and the resulting uneven distribution of temperature within the sample. Intricacies of this kind are of no significance to our story, but should be mentioned in passing just in case the reader's practical experience appears to challenge our description of events.

As the liquid is cooled, and before it reaches its freezing temperature, many other physical properties are gradually changing within it. One of these—the most important from our particular point of view at the moment—is the viscosity. As the temperature falls, the liquid gradually becomes more sluggish in its movement. Nevertheless, for most liquids, the viscosity at the freezing temperature is still small enough to enable the atoms to rearrange themselves into their crystalline pattern in a fraction of a second, and a crystalline solid results. For some systems, on the other hand, the viscosity at the freezing temperature is much larger; so large in fact that the liquid doesn't have time to rearrange its atoms significantly before the temperature moves to lower values for which the viscosity is even larger. Moreover the viscosity increases so rapidly with falling temperature in this region that, at or near a lower temperature called the 'glass temperature', the time required for a rearrangement of atoms to the crystalline pattern becomes (from a practical point of view) essentially infinite. The resulting material is rigid, like a solid, but atomically it has the structure of a liquid that has been frozen in time. It is a glass. In principle, it should transform to a crystal if you could wait forever for the event, but in practice it is quite stable at room temperature for months, years, and often centuries. In fact, theoretically, for many cases it should be quite stable for times exceeding the age of the universe.

According to this picture, it ought to be possible to induce glass formation in any substance if it can be cooled from the liquid state sufficiently quickly. In practice, of course, there are limits to how fast a cooling rate can be attained and it is in general difficult to make glasses except for certain classes of material. However, among these, some are such good glass formers that it is actually more difficult to obtain the crystalline phase than the glassy one, often requiring cooling rates slower than a few degrees per hour. At the other end of the practical glass-forming scale, cooling rates faster than degrees per microsecond are necessary, which can be achieved by forcing the liquid through a small orifice onto and squeezed between massive and rapidly spinning cold metal drums. Of the many physical and chemical properties that determine whether or not a particular material will be a good glass former, the one of primary interest to us (from a mathematical standpoint) is the simplicity of its crystal structure. If the smallest Haüy-type building block, or unit cell, contains a very large number of atoms, or if several different crystal structures are almost equally stable, then the degree of what I call 'confusion' confronting the material at its freezing point is excessive, and the time required for it to 'solve' its crystallization problem is likely to be significant. These kinds of materials do not necessarily have a complicated chemical formula—the confusion can be produced by angular constraints between the atomic bonds that are required by quantum chemistry but which do not enable the atoms to pack nicely into small repeating units. In fact the best known and most widely used glasses (e.g. for windows, bottles, lenses, etc) are all based on the mineral silica, which possesses an extremely simple chemical formula, SiO_2.

Silica forms a glass for both the above reasons. It has several crystalline forms that are closely equal in energy, and it is also angularly constrained such that the Si–O–Si bridge angle does not allow a simple repeating unit of atoms to be formed with only one or two silicon atoms in it. Although first produced as a glass by man over four thousand years ago, its truly detailed local structure in the glassy state remains less than precisely known to this day. In fact, as recently as 1932, W Zachariasen (who is considered by many to be the father of the science of glass structure) could make the statement that 'it must be frankly admitted that we know practically nothing about the atomic arrangement in any glass'. The fundamental problem is, of course, that while crystallinity (or periodicity) can be produced among atoms in only very selective ways based on the Bravais lattices of figure 5, disorder (even when frozen in time) can appear in an infinite number of guises. Just as balls shaken into a container never rearrange themselves twice in exactly the same fashion, so no two glasses are ever identical in structure on the atomic scale. Nevertheless, once again like the balls in a container, certain aspects of randomness (such as packing fraction and density) are very precisely defined. But how far can we carry the analogy? Do the atoms in glasses really pack together like balls? Can the structure of glass teach us anything about how to close-pack spheres in a random manner? Does nature, perhaps, have the answer to the question concerning the most efficient way of packing together spheres in three dimensions without periodicity?

Unfortunately for our ball-packing analogy, the best glass formers consist of a random packing of molecular rather than atomic units. Thus, for example, silica has as its basic building unit the chemically-very-stable tetrahedral molecular configuration SiO_4, with silicon (Si) in the center and the four oxygens at the tetrahedral corners. This unit is rigidly preserved throughout virtually all silica-based crystals and glasses for chemical bonding reasons beyond the scope of this book. Nevertheless, it is a regular tetrahedron, so that we seem at first glance to have 'got lucky' since, after all, we did start this chapter by expressing concern for the space-filling properties of just such regular tetrahedra. Unfortunately, however, from the point of view of dense packing, it turns out that silica has no relevance, since dense packing would require assembling these tetrahedra face-to-face whereas the chemical formula SiO_2 makes it clear that in this material (whether crystalline or glass) the SiO_4 units must share corners. Only in this manner, with each oxygen being shared between two tetrahedral units, can the bulk compositional ratio of one to two between silicon and oxygen be maintained. The glass structure of silica is consequently a rather open one, involving an assembly of tetrahedra all joined by their corners and with rather large spaces between. This is an example of what is known as a 'random network' glass, and has nothing to do with the problem of dense random packing. A schematic two-dimensional version of an 'open' system of this kind is shown in figure 8.

What we need for a physical manifestation of the dense packing of spheres is a glass formed from an elemental material (that is, one possessing only a single species of atom) and one which prefers a dense ordered packing, such as cubic or hexagonal close packing, in its crystalline phase. Unfortunately, from what we have said above concerning the attributes expected of good glass formers, such a material is likely to be very difficult to stabilize in a glassy form. Possible candidates can be found among what are termed the 'transition' metals and the 'noble' metals, specific examples being iron, cobalt, nickel, copper, silver and palladium. Although the cooling rates required to produce

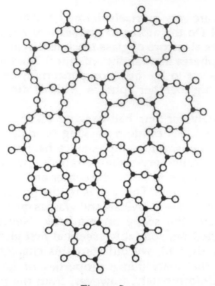

Figure 8

them in non-crystalline form are generally beyond those readily attainable in practice, they can be prepared by other rather special methods. And although they are not very stable, and often tend to crystallize in hours rather than centuries, these elemental glasses do retain their structure long enough for experiments to be carried out upon them. Much more stable glassy metals can be prepared by alloying (that is mixing) with small amounts of other non-metallic elements like phosphorus, carbon or silicon. In proportions that are typically 80% metal to 20% non-metal many of the resulting 'metal–metalloid' glasses can often be formed by direct rapid cooling and are, in many cases, quite stable even to temperatures above room temperature. Some are valuable device materials, being able to display strong magnetism. But our interest concerns the manner in which the atomic 'spheres' pack together in these systems and how closely they resemble a dense packing of hard spheres. In this context these more stable glasses must be modeled in terms of spheres with two different sizes. Our primary interest will therefore still center on the structure of their conceptually simpler, though less stable, elemental cousins. Questions of this kind are most frequently probed experimentally by x-rays.

X-rays are beams of radiation that travel at the speed of light and consist of electric and magnetic fields oscillating in directions perpendicular to the beam itself. They belong to the same family of radiation as light and differ from the latter only by having much shorter wavelengths and higher frequencies. This x-ray wavelength is close to the distance between atoms in solids and it is this particular attribute that makes them ideal for the purpose of determining atomic structures. When the x-ray enters a solid, for example, it encounters the individual atoms. In passing, its oscillating electric field stimulates the electrons in the atoms to emit oscillations in all directions at the same frequency as the x-rays themselves. These 'secondary' wavelets mix with each other (either strengthening or weakening the resultant in any particular direction) in a manner that depends on the positions of the atoms in the solid. If the solid is a crystal, then certain directions exist for which the secondary wavelets strongly reinforce one another, and others for which they cancel out exactly due to the precise periodic arrangement of atoms in crystals. The scattered radiation pattern therefore contains, in a sort of coded form, all the information about the arrangement of atoms in the crystal, and even the actual interatomic distances involved (which are typically, for nearest-neighbor atoms, of order 10^{-8} cm).

The major key for decoding these x-ray 'interference', or 'diffraction', patterns in crystals is exactly the mathematics that led to the proof of the existence of the 14 Bravais space lattices of figure 5. But knowing the appropriate Bravais lattice is only the beginning, since it tells us nothing about the additional atoms that may exist inside the Bravais 'unit cells'. The keys required for unraveling these further details concern the other symmetry operations (in addition to translation) which also leave the crystal structure unchanged (or 'invariant' to use the proper word). These might include reflections in planes, inversions through points, reflections combined with translations, etc. But these ideas, too, had also been studied (and the details perfected) well before the discovery of x-ray diffraction. They developed gradually in the half-century following the work of Bravais himself and are an example of one of the most remarkable instances of independent discovery on record. If one includes all possible symmetry operations for crystals there are no less than 230 different kinds of symmetrically distinguishable arrangements of atoms possible in crystal structures. Evgraf Fedorov (1853–1919), a celebrated Russian crystallographer and mineralogist, began to publish his results in 1885 and completed them by 1890. However, as he wrote in Russian, his work was not at first noticed by mathematicians of the Western World. The German mathematician Arthur Schoenflies (1853–1928) published his work in 1891, and three years later the British scientist William Barlow (1845–1934), who professionally was a London businessman and dabbled in the mathematics of crystal structures only as an avocation, also independently announced his conclusions. Each approached the problem from a different point of view, and all three arrived at exactly the same results, namely the identity and symmetry properties of all 230 'space groups' as they are now called.

The interference of x-rays with crystals, coupled with the mathematical framework concerning symmetry which provided the key to the interpretation of the x-ray interference patterns, provided an extremely powerful weapon for investigating the structure of crystals at the atomic level. Beginning in earnest in the immediate post World War I years, determinations of crystal structures were published in ever increasing numbers. Since large pieces of good quality crystal were not necessary, progress was rapid. In fact, the x-ray method is so powerful that structural decipherment can even be achieved with powders made up of tiny randomly oriented crystallites. The interference patterns for the powder case, unlike the arrangement of sharp dots which characterize crystals, appear in the form of sharp rings around the direction of the incident beam.

The pattern for glasses, although bearing a superficial resemblance to that for a powder, contains fuzzy halo-like rings that increasingly overlap at larger radii until all detail becomes lost. This kind of x-ray pattern is in many respects similar to that observed for liquids and it is the increasing broadness and overlapping of the rings that confirms the non-crystallinity and prevents the determination of the detailed structure. Nevertheless, the inner rings are well defined and do contain important information concerning the nearest-neighbor coordinations of the atoms composing the glass. More precisely, they measure a probability $p(r)$ defined such that the mean number of atoms inside a spherical shell of thickness t and radius r about any particular atom (chosen arbitrarily to be at $r = 0$) is $p(r)t$. Thus, if the glass had no discrete atomic structure at all, like a jelly, then $p(r)$ would just be equal to $4\pi r^2$. More generally, for a real glass, we write

$$p(r) = 4\pi r^2 g(r)$$

in which $g(r)$, which would be equal to one at all distances r for atomic jelly, is called the radial distribution function, and monitors the glass structure to the extent that x-ray diffraction is able to perform the task.

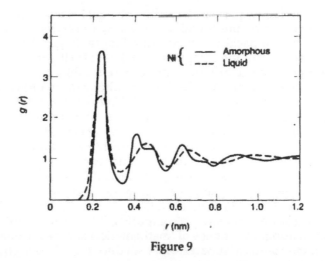

Figure 9

And what does $g(r)$ look like for a real glassy (or some prefer to call it 'amorphous') elemental metal? Well, we show in figure 9 the experimental finding for amorphous nickel, out to a radial distance of 1.2 nm, where a nm, short for nanometer, is 10^{-9} m (a billionth of a meter). On this same figure we also show $g(r)$ for liquid nickel, obtained above the melting point of the ordered metal. The similarity between the two is striking and confirms in a convincing fashion our earlier notion of a glass (whatever its structure) being a time-frozen liquid.

It is not difficult to interpret the $g(r)$ of figure 9 at least in a qualitative fashion. At very small r-values, $g(r)$ is zero. This merely confirms the wholly unsurprising fact that the centers of two nickel atoms cannot approach each other more closely than the sum of their atomic radii (which is a distance of slightly more than 0.2 nm). At r-values a little larger than 0.2 nm we find a large narrow peak. This is produced by those nearest-neighbor atoms which are in contact with each other. Thus, the height of (or more accurately the area beneath) this peak reveals how many 'contact' nearest neighbors the average atom has. The more distant ripples record the second- and third-nearest-neighbor 'shells' of packed atoms, to the extent that this language retains a well defined meaning in the glass structure. Finally, at large r, the distribution function approaches its 'jelly' value of one, the glass structure becoming more and more featureless on average in this limit.

Our interest is centered upon whether this experimentally determined $g(r)$ is anything like that which would result from a dense randomly packed assembly of equal-sized spheres. In particular, has nature found a clever way of randomly packing spheres with a packing fraction significantly larger than that of the ball shakers or the computer modelers? Has it perhaps found a packing more dense than the ordered cubic packing—that mystical random arrangement that mathematicians have never managed to establish as impossible? Unfortunately not! Sad to say, nature has not managed to pack its atoms randomly in a fashion any more efficient than can be achieved by the ball shakers. This fact can be deduced by direct computation of $g(r)$ for balls poured into large containers (and persistently shaken in order to maximize the density) although it is far from trivial to carry out the counting and measuring necessary to calculate $g(r)$ for this case. Far simpler, nowadays, is to resort to computer modeling since, with today's computers, it is not at all difficult to 'build' random assemblies of 'mathematical spheres' containing tens of thousands of spheres, and to deduce the relevant $g(r)$ with great precision without ever leaving the computer terminal. It all sounds easy, but some thought is required since it is, of course, necessary to invent a rule (or 'algorithm' as mathematicians prefer to call it) for building up the assembly, and no rule presents itself as being obviously the most efficient for maximizing the packing fraction.

The simplest one is to start with a 'seed cluster' of four spheres, each touching the other three, and to choose some arbitrary point inside the cluster as the 'origin' or center of the assembly to be built. Spheres are then added sequentially, each to touch three members of the existing cluster, and in such a manner that each new addition is placed as close as possible to the arbitrarily chosen origin. For different choices of origin the final aggregates of thousands of spheres are, of course, all different in detail. They all, however, give essentially identical radial distribution functions $g(r)$ which do indeed mimic the experimental curve of figure 9 rather well. The agreement with experiment can be improved further by replacing hard spheres with 'soft' ones (where soft spheres allow for minor deformations in shape to occur if the latter increases the packing fraction f) suggesting strongly that real elemental amorphous metals are very close physical manifestations of randomly packed spheres—but packed no more cleverly than balls thrown into a large bucket.

There are, of course, other computer algorithms for preparing dense random assemblies of equal sized spheres, and each vies with the others to achieve the largest packing fraction (or equivalently density). The record to date, however, still hovers at or just below $f = 0.65$ (not significantly larger than the 0.64 value cited earlier for balls thrown into a can), and they all produce $g(r)$ curves in close agreement with those experimentally measured for amorphous elemental metals. Apparently nature itself seems unable to find a dense random packing arrangement for spheres with f larger than 0.65. You may argue that this is perhaps because, in the ultrafast deposition methods necessary to make these amorphous metals, we just do not give nature long enough to 'find' that very special dense random packing with f larger than the densest ordered packing (with $f = 0.740\,48$). But if we do give it this extra time, then it always forms the ordered lattice anyway, as if in desperation! If only Aristotle had been correct in his assumption of the space-filling properties of tetrahedra then none of these random packing problems would arise. These amorphous elemental metals would quickly form the ultimately dense perfect tetrahedral packing with $f = 0.7796\ldots$. But Aristotle was mistaken and they cannot. So you see now why all scientists except mathematicians are quite happy with the notion that the densest possible packing of spheres in three dimensions is that of the ordered arrangement of cannon balls in front of the War Memorial. Being a physicist by trade myself, I am happy to go along with them—although (recalling the anecdote in chapter 1) I must admit that I do still keep a watch out for that sheep with one black side and one of some other color!

3

From Euclid to General Relativity

The origins of geometry are lost in the mists of time. However, it is clear that by the era of the great Greek Philosophers, a fairly wide range of geometric knowledge had accumulated and was finally formalized in Euclid's monumental work entitled *Elements*. Considering his fame, remarkably little is known of Euclid's personal life—not even his dates of birth or death, although one often sees the span 300–275 BC quoted. It is clear that after the death in 323 BC of Alexander the Great, whose empire stretched from Greece and Egypt to the borders of India, there was strife among the generals in his army. Leaving his empire, in his own words, 'to the strongest', it is not surprising that the decades following his death saw many dire conflicts and, by 306 BC the Egyptian portion of his empire was in the hands of Ptolemy I. Ptolemy, who as a general in Alexander's army had taken a leading part in the later campaigns in Asia, assumed the title of king and made Alexandria his capital. It was there that he established an Institute to which he attracted many of the leading scholars of the day. Among them was Euclid, who became so associated with the school that he is often referred to as Euclid of Alexandria.

Euclid and his *Elements* are often regarded as synonymous although, in fact, Euclid authored many other texts on widely varying topics in mathematics, physics, astronomy, and even music. His part in *Elements* appears to have been primarily that of a coordinator and editorial arranger of existing geometrical knowledge. Most importantly, he provided a logical development of geometry such that every statement could be referred back to a collection of 'postulates'. These postulates, or 'axioms', are a class of propositions that are simply asserted to be true.

Euclid, ca 300 BC. Reproduced by permission of Mary Evans Picture Library.

They cannot be proved since, without the postulates themselves, there is no framework within which to prove anything. They were, however, to Euclid, more than a mere set of non-contradictory statements. They were intended to be assertions about the real geometric properties of the actual space in which we live, and to be so self-evidently true that no-one could possibly question them. From them, all the remaining 'less than self-evident' properties of geometry were to be deduced by pure logical reasoning alone.

Since this Euclidean teaching was to dominate geometrical thinking and instruction for over 2200 years (and even, to this day, plays a dominant role in High School text books on the subject) it is important for us to look closely at the five simple postulates upon which the entire structure is founded. The first four can be stated quite concisely as follows:

1. There is exactly one straight line connecting any two distinct points.

2. A straight line may be extended in a straight line in either direction.

3. About any point a circle of any specified radius exists.
4. All right angles are equal.

Implicit in these assertions are the 'facts' that space is unbounded, continuous, and homogeneous (in the sense of possessing properties which are independent of direction). The fifth and most famous (or infamous) postulate is rather cumbersome when stated in its original form as follows:

5. If a straight line falling across two straight lines makes the sum of the interior angles on the same side less than two right angles, then the two straight lines intersect, if sufficiently extended, on that side.

This implies, in particular, that if two lines (and lines, in the present context, always mean straight lines) are crossed by a third which makes a set of right angles exactly with each, then the two original lines never meet. Since, by definition, straight lines that never meet are called 'parallel', the fifth postulate can be much more concisely stated using this notion of parallelism. In this form, attributed to the Scottish mathematician John Playfair (1748–1819), it is most commonly referred to as Euclid's Parallel Postulate:

5. Through a point outside a given line, one and only one line can be drawn parallel to the given line.

Providing that it is made quite clear what is meant by a straight line (namely the shortest distance between two points), a right angle (the angle defined when two straight lines cross in such a manner that the four angles so defined are all equal), and a circle (the locus of all points equidistant from a point), then it is indeed quite difficult to find anything to quarrel with in the first four postulates. The fifth, however, includes (explicitly in its original form) the phrase 'if sufficiently extended'. This carries with it the possibility of extension to arbitrarily large distances, and therefore the notion that we know something precise about the behavior of 'real' space over arbitrarily large distances (in addition to its assumed continuity and homogeneity). It was this fact that was, over many later centuries, to cause great concern and to be the focus of immense controversy. Nevertheless, the fifth postulate is absolutely essential for establishing most of the geometric results that we remember from High School geometry classes including, for example, the finding that the angles of a triangle add up to 180 degrees (or two right angles, as Euclid would have stated it).

It is true that certain additional non-geometric axioms are also required before the construction of the enormous edifice of Euclidean geometry can begin: axioms like 'things equal to the same thing are equal to each other' and 'a whole is greater than any of its parts'. But none of these would be likely to raise any eyebrows (at least until the challenge of classifying the truly infinite was taken up in the latter part of the nineteenth century) and, judged by the brevity and smallness in number of the 'self-evident' assumptions, the range and varieties of the theorems that follow from them is quite astounding. In Euclid's highly acclaimed *Elements* (which was actually divided into 13 books) no less than 465 propositions were deduced, including examples from both plane and solid geometry. Although it is generally agreed that few of these theorems originated with Euclid himself, his superbly organized treatise was so revered that it completely obliterated all preceding works of its kind and, although some modern-day geometers have questioned the completeness of some of the proofs, not a single one of the 465 propositions is false.

We shall, in passing, cite only one of these 465 propositions — namely Euclid's proof of the Pythagorean theorem, since it is very different from that normally found in textbooks today. The modern approach uses similar triangles and simple proportion which Euclid did not wish to introduce until later (Book VI to be precise) in his *Elements*. However, he wanted to reach the Pythagorean theorem as early as Book I, and he thus devised a proof (most now believe that it is original with Euclid) of stunning simplicity. In its avoidance of proportions it also avoids the 'uncomfortable' problems concerning irrational numbers which the more modern proofs must confront (an irrational number being one which cannot be expressed as the ratio of two whole numbers or, if you prefer, has a decimal form that goes on forever without repeating — like $\sqrt{2}$ or $\sqrt{3}$).

Euclid's proof of the familiar 'square on the hypotenuse is equal to the sum of the squares on the other two sides' is easily appreciated with reference to the right-angled triangle and its associated squares shown in figure 10. The square on side AC is equal to twice the area of the triangle FAB (same base, equal heights) or to twice the triangle CAD (congruent triangles) or to the rectangle AL (same base and same height as triangle CAD). In a similar fashion, the square on BC is equal to twice the triangle ABK, or to twice the triangle BCE or to the rectangle BL. Hence, the sum of the squares is equal to the sum of the rectangles, which is the square on the hypotenuse AB.

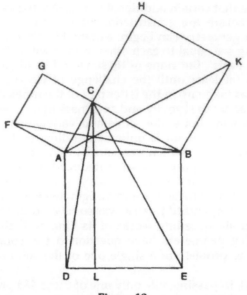

Figure 10

However, it is not the purpose of this chapter to spell out elegant examples of Euclidean geometry. What it is important to realize is the fact that, remarkable though the Euclidean system is, it is not (even in principle) able to establish *all* geometric truths. In fact, it is not adequate to prove some that were already known to the early Greeks themselves. A spiral, for example, with a modern polar coordinate representation $r = a\theta$, just cannot be constructed with the straight edge and compasses required for Euclidean arguments. Geometric truths concerning curves of this kind depend, for their proof, upon certain 'limit' operations that were only fully perfected after the invention of differential calculus by Isaac Newton (1642–1727) and Gottfried von Leibniz (1646–1716); see chapter 12. In spite of this, even as early as the mid-third century BC, the Greek mathematician and inventor Archimedes (287–212 BC) was familiar with many properties of the spiral. Thinking of a point on the spiral $r = a\theta$ as resulting from a double motion (one moving uniformly along the ray r while the other θ moves the ray itself uniformly about its fixed end) he constructed tangents to the curve using methods that came very close indeed to differential calculus itself.

Even with the recognition that certain geometric problems (for example, all truths about figures constructed on curved surfaces) are outside the realm of pure Euclidean geometry, the power of the Euclidean method remains impressive. Nevertheless, almost from the very beginning, there was an uneasy feeling about the fifth postulate—that it was, in some sense, less self-evident than the others. Among many mathematicians and geometers over the centuries there was a suspicion that possibly the fifth postulate was not an independent statement at all; that, perhaps, it could somehow be proved from the other four. Scores of erroneous 'proofs' appeared over the years, always to be finally demolished upon closer examination. Even such an accomplished mathematician as Joseph Louis Lagrange (1736–1813), who is largely credited with systematizing the field of differential equations, was included in their number, although it is said that during a presentation of his 'proof' to the French Academy he recognized the flaw in his argument and retired from the hall in embarrassment.

Perhaps the most famous of all such attempts was that of the Italian mathematician Giovanni Saccheri (1667–1733). In the very year in which he died, he published a book entitled *Euclid Freed from Every Flaw* in which he made an elaborate effort to prove the parallel postulate. From each end of a baseline, and on the same side, he raised two lines of equal length at right angles to the base and then joined them at the top. To us, his construction would look like a rectangle. Nevertheless, without the parallel postulate it is not possible to prove that the resulting quadrilateral contains four right angles. Although the two angles involving the baseline are right angles by definition, the other two, without the fifth postulate, can only be shown to be equal. By considering in turn the three possibilities that the unknown angles are acute, obtuse, or right angles, he derived theorem after theorem without ever encountering any difficulties. In fact, we now know that he was building up the first perfectly consistent non-Euclidean geometries. But his conviction, that the fifth postulate somehow just *had* to be true, eventually persuaded him to reject all the non-Euclidean findings by what he considered to be a *reductio ad absurdum* method. Where no contradictions really existed he twisted his reasoning to interfere with his logic. And so it was that, after all his earlier highly commendable work, he ended up producing just another flawed 'proof' of the fifth postulate as derived from the other four. Thereby, he lost all credit for what would have been the most significant mathematical advance of the eighteenth century—namely the discovery of non-Euclidean geometry.

As it was, efforts to prove the parallel postulate continued well into the nineteenth century. Early in the nineteenth century, three mathematicians almost simultaneously had the burst of insight necessary to see the matter in its true light. The first of these was the incomparable Gauss who, wanting to prove that the angles of a triangle must add up to 180 degrees assumed, for the sake of argument, that they did not. Beginning with an assumption that the required sum is less than 180 degrees, Gauss derived a seemingly bizarre geometry which, however, appeared to contain no logical inconsistencies. The more he pursued it, the more he began to sense that this new geometry was a true alternative to that of Euclid. He said as much in a private letter in 1824. And yet, for some reason, he failed to publish any of his findings. Some feel that, being already the foremost mathematician of his day, he feared that the controversial nature of the work would cause an uproar that might jeopardize his reputation.

The next player in this saga was the young Russian Nikolai Ivanovich Lobachevsky (1793–1856). He, like Saccheri and Gauss, set out initially to prove the parallel postulate. In this endeavor, he not only convinced himself that the fifth postulate could not be derived from the other four, but that it is not necessary (and may not even be true) in any strict sense. Lacking the diffidence of Gauss in this matter, Lobachevsky published his results for all the world to see. The year was 1829, and in his first article on the subject he created a whole new and self-consistent geometry that did not contain the parallel postulate. He thus became, in a sense, the 'Copernicus of geometry', liberating the subject from the shackles imposed by the parallel postulate. It was not that Lobachevsky denied that Euclidean geometry seemed to work admirably in the real world around him in a practical sense, what he did establish was the fact that an equally non-contradictory geometry could be constructed with a fifth postulate completely different from Euclid's choice. Indeed, Lobachevsky referred to it as an 'imaginary' geometry—but, as we shall see, it was actually far from that.

Lobachevsky adopted as his fifth postulate the statement that 'through any point outside a given line, more than one line can be drawn parallel to any given line'. From his new geometry could be drawn many bizarre and distinctly non-Euclidean findings such as:

(1) the interior angles of a triangle add up to *less* than two right angles;
(2) the ratio of the circumference of a circle to its radius is *greater* than two pi;
(3) no similar figures of different areas can exist.

With such a range of results seemingly so in conflict with everyday experience, it is perhaps not surprising that the new geometry was slow to be accepted by others. But it turns out that Lobachevsky had not been alone in coming to a realization that non-Euclidean geometries had merit.

Next on the scene was the Hungarian mathematician Janos Bolyai (1802–1860). Janos' father, Farkas, who had been an associate of Gauss, had himself spent much of his life in a futile attempt to prove the parallel postulate. As a result of his experiences with the problem he warned his son to stay away from it. But Janos did not heed the advice and, in work which in many ways parallels that of Lobachevsky, he also finally reached the conclusion that Euclidean geometry had a perfectly viable competitor. He, too, was amazed by its bizarre yet apparently quite consistent propositions and wrote 'Out of nothing, I have created a strange new universe'. He published his work in 1832 in a 29-page appendix to two thick volumes of rigorous expositions of geometry, arithmetic and algebra authored by his father. It was a most concise presentation of what he called his 'absolute geometry'. The elder Bolyai enthusiastically sent a copy to his friend Gauss who very ungraciously downplayed the findings by claiming that the entire content merely coincided with his own 'meditations which have occupied my mind for from 30 to 35 years'. On top of this disappointing reception of his work, Janos was soon to find out that Lobachevsky had published similar findings some three years earlier. Lobachevsky, however, had written in Russian and his work had gone unnoticed in Western Europe.

The new non-Euclidean geometry remained as a fringe aspect of mathematics for another decade or two until the problem was taken up afresh by the German mathematician Georg Friedrich Bernhard Riemann (1826–1866). Gauss, Lobachevsky and Bolyai had all focused on the geometry for which the sum of the interior angles of a triangle is less than 180 degrees. This had been done for a good reason. The opposite possibility, of having the angles of a triangle add up to more than 180 degrees, had led to a logical contradiction in the limit that the sides of the triangle were allowed to become infinitely long. But was there a need to consider this limit at all? Riemann said no. Euclid's second postulate stated merely that straight lines could always be continued, and this was not the same as asserting that they had to be of infinite length. Riemann could easily imagine geometries where lines, somewhat like circles, are of finite length but yet have no end. When Riemann examined geometry under the assumption of finite yet unbounded lines he established that, once again, a perfectly self-consistent form could be defined with

Georg Friedrich Bernhard Riemann, 1826–1866. Reproduced
by permission of AIP Emilio Segrè Visual Archives.

yet another fifth postulate: namely that 'through any point
outside a given line, *no* line can be drawn parallel to the given
line'. From this new non-Euclidean geometry one could establish
such unlikely results as:

(1) the interior angles of a triangle add up to *more* than two
right angles;

(2) the ratio of the circumference of a circle to its radius is *less*
than two pi;

(3) no similar figures of different areas can exist;

(4) a straight line can always be continued but is never of
infinite length.

Now all of this, in spite of its claimed self-consistency, must
have had a slightly 'Alice in Wonderland' appearance to the
scientists of the day, even if any were aware of it (and anyway,
Alice had yet to make her debut in literature). And yet, perhaps
somewhat surprisingly, it is not at all difficult to envisage a
universe in which Riemann's geometry, and not Euclid's, is valid.

This is most easily accomplished by stepping down for a moment from three dimensions to two. If, for example, the universe were two-dimensional in its local appearance (rather than three), then two-dimensional Euclidean geometry (complete with its five postulates) could be established for it in a manner that exactly parallels the formulation of the Ancient Greek original. In particular, the parallel postulate would be just as problematic in the planar context as it has been in three dimensions.

Imagine, however, the possibility of this two-dimensional universe being really a giant three-dimensional spherical surface, so large in diameter that it appears (to the two-dimensional scientists inhabiting it) to be completely Euclidean within the accuracy of their experiments. What postulates of geometry are actually valid in this new domain? Straight lines (being the shortest distance between two points) are now arcs of great circles on the universe sphere, like lines of longitude on Earth. In particular, no line can be drawn parallel to a given line, and a straight line can always be continued but is never of infinite length—all in true Riemannian fashion. If you picture yourself as a god, who can exist outside of this universe, then to you the universe described above may be no more than the surface of a ball. In your godly role, using a real ball, you will now find it easy to verify some of the other, previously baffling, findings given above for the geometry of Riemann. For example, with one point at a pole and two on the equator, it is not at all difficult to draw a triangle in this strange new world for which each angle is a right angle (figure 11), making the sum of the interior angles of this particular triangle equal to 270 degrees. Also with the center at

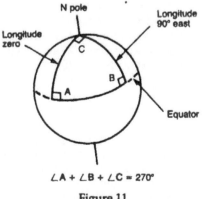

$\angle A + \angle B + \angle C = 270°$

Figure 11

the pole, and a radius equal to the distance between the pole and the equator, we can draw a circle (i.e. the equator) for which the ratio of the circumference to the radius is four (which is, indeed, significantly less than two pi). As a god, all this is easy for you to verify for two reasons. Firstly, you exist in a dimension higher than that which is familiar to the inhabitants of the universe; and secondly, for you the universe in question is small enough to carry out measurements over distances comparable to the size of the universe itself.

Unfortunately, to the two-dimensional inhabitants of the Riemannian spherical world, the third dimension is as incomprehensible as the fourth dimension is to us. In addition, they inhabit only a tiny, seemingly flat, portion of this universe. Nevertheless, with sufficient accuracy of measurement, it should be possible for them to detect the non-Euclidean nature of their environment by carrying out experiments designed to look for small violations of some of Euclid's predictions. And so it is for us. For example, if the universe in which we live were actually a four-dimensional sphere, then its true geometry would be universally Riemannian, although in local regions it would appear to be Euclidean. Clearly, the possibility exists for deviations from Euclidean behavior in our world, and it is the job of experiment to decide.

It is readily established that all the findings of Riemannian 'two-geometry' are true on the surface of a three-dimensional sphere (except at antipodal pairs of points—like the North and South Poles—which violate Euclid's first postulate). For Lobachevskian 'two-geometry' it is a little more difficult to find a simple analogous example, although curves on hyperboloids of revolution (which are hyperbolae rotated about their axes of symmetry) do come close to satisfying the relevant postulates. In a general universe of two, or even three, dimensions, it is even quite conceivable that some parts could tend toward Riemannian 'curvature', while other parts tended toward the Lobachevskian. On the other hand, until experimental accuracy caught up with the challenge, it remained quite possible (in the mid-nineteenth century) that our universe was actually just as Euclidean as everyone had been convinced that it must be up until the non-Euclidean mathematical breakthrough.

Riemann's contributions now went beyond the question of whether zero, one, or more than one, parallel line can be drawn through a point parallel to a given line in geometry. He developed a mathematical formalism capable of defining geometries that are non-Euclidean in a far more general sense, concerning the type and degree of 'non-Euclidean' behavior and its possible variation from point to point in space. To do this, Riemann couched his arguments in terms of analytic geometry and, in particular, of differential forms. In this language (see appendix 1) the distance between two neighboring points (x, y, z) and $(x + dx, y + dy, z + dz)$ in Euclidean space is given by

$$ds^2 = dx^2 + dy^2 + dz^2.$$

In adopting this form Riemann himself was already standing on the shoulders of giants.

There is no unanimity of opinion among historians concerning who first 'translated' geometry into algebra or, to put it in more modern terms, who invented analytic geometry. The concept of fixing the position of a point by means of choosing suitable coordinates certainly goes back as far as Ancient Egypt. The rather more advanced idea that certain algebraic equations, such as $x^2 + y^2 = a^2$, can be interpreted geometrically (specifically in this case as a circle of radius 'a' centered at the origin $x = 0, y = 0$) also goes back at least to the times of the Greeks of the third century BC. The real essence of analytic geometry, however, is the idea that geometric problems can be translated into algebraic equivalents, solved as algebra, and then translated back again to the pictorial representation of geometry. At this level, the most decisive advances were made by the French philosopher and mathematician René Descartes (1596–1650) from whose name we derive the term 'Cartesian' coordinates, and by that prince of amateur number theorists Pierre de Fermat (1601–1665). They tended to approach the problem from opposite directions, with Descartes defining geometrical forms and finding their equations, and Fermat starting with equations and plotting their geometric representations.

After the development of the calculus (see chapter 12) a natural progression took place from the analytic geometry of the finite to that of the infinitesimal limit, and it is in this limit (with ds, dx, dy, dz being infinitesimally small quantities) that the above equation is to be interpreted. Using this language of 'differential' analytic geometry, Riemann soon recognized that such geometries could be defined in spaces with more than three dimensions. It was only necessary to replace the three numbers used to label points in three dimensions by four numbers for four dimensions, five numbers for five dimensions, and so on. Even more importantly, he recognized that the local properties of these spaces could always be uniquely defined by the equation that related ds to its components along the various axes defining the space. Thus, in three dimensions, the equation $ds^2 = dx^2 + dy^2 + dz^2$, if valid for *all* line segments d*s*, defines a three-dimensional Euclidean space. However, infinitely many other formulae can replace this simple Euclidean one in order to define non-Euclidean spaces in three dimensions. For example, a space for which

$$ds^2 = g_{11}dx^2 + g_{12}dxdy + g_{13}dxdz + g_{21}dydx$$
$$+ g_{22}dy^2 + g_{23}dydz + g_{31}dzdx + g_{32}dzdy + g_{33}dz^2$$

where the *g*s are, in general, functions of *x*, *y*, *z*, is now known as a Riemannian space, regardless of whether it corresponds locally to the original Riemannian or Lobachevskian geometries. In particular, a space which is locally Euclidean in some region (with $g_{11} = g_{22} = g_{33} = 1$, and all other *g*s equal to zero in this region) might actually be part of a more general Riemannian space. The pattern, or matrix, of numbers

$$g_{11} \quad g_{12} \quad g_{13}$$
$$g_{21} \quad g_{22} \quad g_{23}$$
$$g_{31} \quad g_{32} \quad g_{33}$$

is called the 'metric' of the space. Most importantly, it was Riemann's suggestion that spaces with different metrics should be the focus of mathematical study that opened up the entire field of modern analytic geometry. Although forms other than that given above could easily have been conjured up to relate d*s* to d*x*, d*y*, and d*z*, the Riemannian 'manifold', defined by the above quadratic form, is the simplest pure generalization of the Euclidean ideal. Thus, it was this form that was the focus of mathematical study in the years following Riemann's death, and it was to just such a representation that Einstein appealed in his development of the theory of general relativity in 1916.

In the real world it is experimentally quite clear that straight lines over short distances satisfy the Euclidean metric $ds^2 = dx^2 + dy^2 + dz^2$. Deviations, if they exist at all, are immeasurably small. Thus, not surprisingly, for the rest of the nineteenth century, Riemannian spaces were at best accepted only as mathematical abstractions. As a philosophy of space they had little effect except to evoke criticism. This came particularly from the German physicist Hermann von Helmholtz (1821–1894) who argued that since freely moving *rigid* bodies could not exist in a Riemannian space for which the *g*s are functions of *x*, *y*, *z*, much of the generality of Riemann's concept must be philosophically unsound. However, we now know that there was more profound wisdom in Riemann's metric than the scientists of the day were able to comprehend. It is, in fact, not possible to construct any body (rigid or otherwise) apart from the space in which it exists. But this realization was still in the future.

In order to test for deviations from the Euclidean nature of space, it is necessary to examine straight lines over very long distances. Physically, a straight line can be defined by the path of any particle or wave which propagates without interference from any external force or field. The most convenient such 'tool' for experimental purposes is the ray of light, since it can cover immense distances in short periods of time. Such rays reach us from the outer regions of space every starry night. The primary problem in experimenting with light over galactic distances is the universal presence of gravity. Two questions immediately arise. First, does gravity affect the motion of light as it travels through space, and second, if it does should we try to subtract out its effects in defining the geometry of the space in which we live? According to Newton's laws of gravity, whether or not light is 'bent' by gravitational forces depends upon whether light is made up of particles or waves, and at the end of the nineteenth century, the answer was not known. In fact, light in its experimental behavior appeared to have a dual 'wave–particle' almost Jekyll and Hyde nature (and still does). But even if light is affected by matter (via gravity), does it make any sense to try to allow for this and to deduce what would be the geometry of an empty universe? In an empty universe there would be nothing to measure with and nothing to measure with reference to—no position, no velocity and, most importantly, no acceleration. For example, would a lone bucket of water in an otherwise empty universe 'know' whether or not to 'slop-over'? At the dawn of the twentieth century all this appeared too baffling to contemplate. The time was ripe for an Einstein.

Albert Einstein (1879–1955) was born in Germany and educated primarily in Switzerland. The young Einstein was far from being a mathematical prodigy. As a child he seemed intellectually backward and he failed his first entrance examination to Zürich Polytechnic where he wanted to study to become an electrical engineer. After an extra year of preparation he finally obtained entrance to Zürich (now, however, intending to train as a physics teacher). He didn't adapt well to the regimentation imposed by the system of examinations (which constrained his mind) but he did pass his final examination—though not spectacularly. Zürich didn't keep him on, even as an assistant—its lowest grade of post-graduate employment—and Einstein found it necessary to search for employment outside of the academic world. He

secured a position as a patent examiner at the Swiss Federal
Patent Office. Here he found that he could accomplish his patent
work without undue effort and he soon became financially
secure. Thus, at age 23 (in 1902) he finally had time to work on his
doctorate and, more importantly, to meditate on the nature of the
physical universe. Shortly thereafter (1905) Einstein published
the first of his several papers on relativity—papers that would
revolutionize man's image of the physical universe.

The earlier papers concerned what is now known as the
'special theory' of relativity. This theory, with all its implications,
has been thoroughly tested in countless contexts throughout the
twentieth century and is now accepted by almost all concerned
with the subject. In fact, to most physicists today it is counted as
one of the two cornerstones of modern physics—the other being
quantum mechanics (see chapter 12). The theory arose out of
contradictions between the disciplines of electromagnetism
(which is the interplay of electricity and magnetism) on the one
hand, and Newton's laws of motion on the other. At the root of
the problem is the concept of time. To Newton, in his *Principia* of
1686, time is 'absolute, true, and flows equably without relation to
anything external'. Einstein found himself forced to question this
for the following reasons. He was able to formulate two postulates, each of which appeared to have solid experimental support,
but which together could not be reconciled with Newton's
inexorable concept of time. The first was the 'Principle of Relativity' which stipulated that the laws of physics (and in particular of
electromagnetism) are the same for all 'inertial' (that is, not
accelerating) frames of reference. Put another way, this states
that there is no experiment capable of detecting *absolute* velocity;
only the *relative* motion of different 'observers' can be measured.
The second postulate stipulated that light always moves with the
same velocity in any inertial frame of reference. If both postulates
were true (and the experimental evidence was overwhelming)
then space and time could not possibly be independent quantities. They had to become relative concepts, fundamentally
dependent on the observer.

In such a fashion, the concept of 'spacetime' entered physics.
Thus, whereas in Newtonian physics each observer could be
labeled by his location in space (x_i, y_i, z_i) together with a universal
time t that was the same for each one of them, in Einstein's world
each observer needed to possess his own unique measure of time
t_i as well. In this case, each observer is now 'located' by the
spacetime coordinates (x_i, y_i, z_i, t_i). From this requirement of the
'special theory', many weird and wonderful predictions follow,

including the variations of length, mass, and time intervals with velocity. All these effects involve corrections to the 'common sense' notions of Newton only of order v^2/c^2, where v is the velocity of the object in question relative to the observer, and c is the velocity of light. Since this ratio is extremely small for the kind of velocities that we commonly observe about us, the peculiar results of Einstein's theory are not immediately apparent to our senses. But they have all received accurate verification by direct experiment, and in some cases important to experimental physics the effects are not small at all. For example, in experimental high-energy physics (in which electrons and other particles are accelerated to velocities quite close to c) the design of the accelerator itself relies overwhelmingly upon the predictions and calculations of the special theory of relativity. The same accelerator also enables physicists to verify one other peculiar feature of the equations of the special theory—the fact that no particle can be accelerated to velocities exceeding the velocity of light.

The difference between the Newtonian world and its Einsteinian counterpart is perhaps best illustrated by diagrams (figures 12(a), (b)). In these figures, for simplicity, we measure 'space' in only one dimension. Then, instead of spacetime being a four-dimensional concept, it reduces to a two-dimensional situation that we can draw. In Newton's world, if we plot spatial position on the horizontal axis and time on the vertical axis, then lines of 'simultaneity' (for which all observers agree that the time has the same value for each) are simply horizontal lines (figure 12(a)). In Einstein's world of special relativity this spacetime universe is still two-dimensional, but observers at different space points x now measure time differently. Each point in this space can be referred to as an 'event', but there is no preferred way of assigning separate space and time coordinates to each event. Indeed, observers at different spatial positions x may well not agree as to when two events are simultaneous. Let us ponder this strange situation in a little more detail.

When there is a light ray connecting two events, then and only then is there no doubt concerning simultaneity, since light travels at the same speed (close to a billion miles an hour) regardless of the relative velocities of the observers or events. Thus, for an arbitrary observer, whom we place for simplicity at the origin of our spacetime coordinate system, the lines measuring simultaneous events in his or her frame of measurement (x,t) are just the lines $x^2 = c^2 t^2$ (figure 12(b)). If the event (x,t) is such that $\sqrt{(c^2 t^2 - x^2)}$ is real and positive, then (for this and all other observers) it is in the future. Correspondingly, if this same

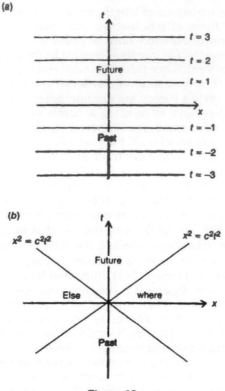

Figure 12

quantity is real and negative, then the event for all observers is in
the past (figure 12(b)). But what about all the events on the x,t
plane that are 'elsewhere'? These events are not absolutely in the
past, present or future—different observers will come to different
conclusions concerning them. On the other hand, for these there
is nothing the observer can do to affect the event, and no way that
the event can affect the observer at the time of observation, since
no 'causal' effect can travel faster than light. The importance of
the quantity $c^2t^2 - x^2$ in all of this is the fact that all the observers
(at different points in spacetime) will agree as to its value, but will
disagree concerning the separate time and spatial values x and t.

Unfortunately this is not the place for us to probe the fascinating consequences of these unfamiliar notions. It is sufficient for
our present story to note that the lines $x^2 = c^2t^2$, which become

cones in three-dimensional spacetime (x, y, t) and which we may still refer to as 'cones' in the 'real' four-dimensional spacetime (x, y, z, t) of our actual local environment, play an essential role in special relativity. In Newton's picture there are 'planes' of absolute simultaneity, which are replaced in Einstein's picture by light cones. If the origin of the light cone in question is placed arbitrarily (for simplicity) at the origin (0, 0, 0, 0) of spacetime, then its equation is

$$x^2 + y^2 + z^2 - c^2t^2 = 0.$$

Since algebraically this cone would reduce to a Newtonian plane $t = 0$ if the velocity of light c could be made infinite, it follows that the world of Einstein and Newton would become identical if the speed of light were truly infinite. Our perception of the world as Newtonian (and our bafflement by most of the notions of the special theory of relativity) is therefore our personal inability to detect with our senses the fact that the velocity of light is not infinite.

The 'distance' between any two neighboring points on the Einsteinean light cone can now be expressed in the form

$$ds^2 = c^2dt^2 - dx^2 - dy^2 - dz^2.$$

Writing $x_0 = ct$, $x_1 = x$, $x_2 = y$, $x_3 = z$ now enables us to rewrite this as

$$ds^2 = dx_0^2 - dx_1^2 - dx_2^2 - dx_3^2,$$

or, in Riemannian nomenclature

$$ds^2 = g_{ij}dx_idx_j$$

where the repeated indices imply a summation over all 16 terms (involving $i = 1, 2, 3, 4$ and $j = 1, 2, 3, 4$) and the metric is such that $g_{00} = 1$, $g_{11} = g_{22} = g_{33} = -1$, and all other $g_{ij} = 0$. Although this describes a four-dimensional space, it is still 'flat' in the Euclidean sense, with light traveling in straight Euclidean lines.

Before 1905 space and time were regarded as Newtonian. However, as early as 1870, the English philosopher William Clifford published some speculations concerning the possibility of space being 'curved'—by which he meant, of course, the familiar three-dimensional space that we appear to have around us. After 1905 it became necessary to raise this question, if at all, in terms of the new Einsteinean context of spacetime; that is in contemplation of possible deviations of the above g_{ij} metric components from their special relativistic values. The question did indeed arise in this context out of an attempt to include acceleration, and in particular acceleration due to gravity, within the greater framework of relativity.

That there was a real problem here was immediately apparent since Newton's theory of gravitation, with its statement that the force of attraction between any two masses m_1 and m_2 separated by a distance r is $F = Gm_1m_2/r^2$, where G is the universal gravitational constant, implied the presence of instantaneous action-at-a-distance. By this we mean that one mass could instantaneously determine the force upon (and acceleration of) another residing at a distance r from it. This contradicted the requirement of the special theory of relativity that no causal effect could travel faster than light.

The fundamental connection between gravity and special relativity came via the principle of 'equivalence'. This stated that there can be no experiment capable of differentiating between 'gravitational' mass (which is the mass entering into the Newtonian force equation $F = Gm_1m_2/r^2$) and 'inertial' mass (which is the mass entering the Newtonian equation $F = ma$, relating force to acceleration). Thus, for example, if you were an observer confined to an empty laboratory somewhere out there in space, and you observed an object passing through the laboratory in a curved trajectory, the equivalence principle says that there is no way for you to determine whether the laboratory is accelerating and the object traveling in a straight line through space, or whether the laboratory has no acceleration but the object is in a gravitational field. In other words, gravitational fields are locally exactly equivalent to the acceleration of coordinate frames. And what about light itself? Clearly, if a light beam crosses the laboratory and the laboratory is accelerating in a direction perpendicular to the light beam, then the path of the light will be curved (because the velocity of light is finite). Put in another way, a gravitational field must bend the path of light, and by an amount which depends on the local gravitational strength at the point of observation.

In order to create a self-consistent description of these events it was now sufficient to allow spacetime to acquire a true Riemannian curvature, and to replace the old concept of a gravitational field completely with a Riemannian metric in which g_{ij} is a function of position in spacetime. In this picture, the gravitational force that acts on any 'body' is nothing more nor less than a manifestation of the underlying nature of spacetime through which it is passing. Objects which are free to move in a gravitational field then follow straight lines (or 'geodesics' as they are usually referred to in this more general context) in a curved Riemannian spacetime.

Most gravitational fields with which we are familiar (i.e. Sun, Moon, Earth, etc) produce only very small Riemannian deviations from Euclidean spacetime. At first glance, this statement seems absurd since we know that a ball thrown into the air possesses a trajectory that deviates enormously from a straight line. But we are thinking here of three-dimensional space curves and not four-dimensional spacetime curves. The important point is that the non-spatial dimension $x_0 = ct$ is, for the sorts of motion we are familiar with, immensely longer than the other three. Thus, if our ball stays aloft for a few seconds, and covers a few tens of feet in the spatial dimensions, it moves no less than several hundred thousand miles in the x_0 dimension (since c is about 186 000 miles per second). The path in spacetime for our ball therefore deviates extremely little from the line it would have described in the absence of gravity.

The magnitude of the Sun's gravitational effect on the planets around it is sufficient to produce a Riemannian deviation from the 'flat' metric of Euclidean space of only about one part in a million. Clearly, experimental verification of the predictions of general relativity in or around the solar system require great experimental accuracy. Since gravity varies from place to place in a manner that depends on the distribution of matter, which itself is not stationary, Einstein realized that a self-consistent theory of general relativity must relate Riemann's metric elements g_{ij} to the motion and distribution of matter via a system of differential equations. The resulting theory, first set out by Einstein in 1916, was capable of making many predictions at odds with Newtonian equivalents, but only in the 1970s would experiments capable of a convincing confirmation of them become available.

The most famous prediction was the magnitude of the bending of light by the Sun's gravity, which was twice the magnitude of the effect predicted by Newton's theory for light 'particles'. The first experimental measurement took place during the solar eclipse of 1919. Although not of great accuracy, it did seem to confirm Einstein's prediction (which was twice the Newtonian one) and was more than sufficient to make Einstein an instant celebrity. Much more recent experimental work has provided extremely accurate confirmation of this and many more of the theory's predictions. Included among them are the measure of gravitational 'redshift' (the change in the frequency of light as it moves through a varying gravitational environment), now verified to 1 part in 10 000; the precession of the orbit of the planet

Mercury, now verified to rather better than one per cent; and the deflection of light by gravity, now accurate to 1 part in 100 000. Clearly, modern-day tests have left little doubt as to the validity of the general theory, at least for objects of macroscopic size and for gravitational fields of the size of those found in our immediate area of space.

All, however, is not perfect. On scales of the order of atomic dimensions or less, the theory does not seem to be compatible with that other bastion of modern physics—quantum mechanics. Also, at the other end of the scale, general relativity remains largely untested in cases for which the Riemannian deviations from the 'flat' metric are truly large (as is the case for 'black holes'). And finally, one extremely important prediction of the theory has, at least at this writing, still evaded detection. It is the existence of the gravitational waves that, within the theory, replace the instantaneous Newtonian interactions between masses. These waves are propagating disturbances of spacetime curvature which travel at the speed of light. Some indirect evidence for their existence has been found from the observation of companion neutron stars, which are losing energy at a rate corresponding to the emission of gravitational 'radiation', but the race for their direct detection as they pass by Earth is one still avidly pursued by experimental physicists (no doubt with a Nobel Prize awaiting the winner).

Thus, it is now clear that our universe exists in a spacetime of four dimensions, and that this spacetime can be represented by a Riemannian manifold with elements g_{ij} $(i, j = 1, 2, 3, 4)$ at each point related to the distribution of matter via a system of differential equations. However, since we do not know the precise distribution of matter throughout the actual universe in which we dwell, the true shape of this spacetime on a universal scale can only be a matter of speculation. In this context, Einstein's equations admit to a variety of solutions of many possible physical geometries, which include among them the notion of an expanding universe of the kind presently believed to have relevance for us. The universe can, of course, have an overall Riemannian curvature in addition to the local 'wrinkles' in the immediate neighborhood of stars and planets—but does it? Apparently it all depends on whether there is enough matter in the universe to reverse the expansion of the 'big bang' or not. The tug of gravity from all the galaxies is certainly slowing down the rate of expansion, and one can actually calculate the critical mass density d_{crit} necessary to accomplish this. That is, if d is greater than d_{crit} (as measured at the present time) then the expansion eventually reverses, and if d is less than this value it expands forever. It can be shown that only if d is exactly equal to d_{crit} is spacetime Euclidean on the grand scale. There is a strong theoretical 'wish' among cosmologists for this to be so, since only in this case does the Big Bang Theory of creation fit simply with the observed size of the visible universe. The required density is only about 10^{-29} g cm^{-3}, but it has so far proven difficult to find even this much among the visible components of the universe, and experimental results remain confused in this context since the likely amount of non-visible matter is still a topic of heated discussion. In spite of these remaining problems, there is no doubt that spacetime in our galaxy is not Euclidean, and that the geometry to be used (if precision is required) should be the non-Euclidean creation of Georg Bernhard Riemann—with predictions so unthinkable that they caused Bolyai to fear for his sanity and Gauss to fear for his reputation.

4

From Plucking Strings to Electrons in Solids

It is known that Pythagoras, as long ago as the sixth century BC, exhibited a great interest in the laws of musical harmony as generated by the plucked strings of the lyre. The lyre usually consisted of a sounding board (often composed of a tortoise shell covered with bull's hide) from which curved horns extended in a U-shape to be connected near their tips by a wooden crosspiece. Strings of gut were connected between the crosspiece and the bottom of the shell and were plucked with one hand and 'stopped' with the other. The pitch of the notes (or sound frequency as we should now refer to it) depended on the length of the vibrating string, and Pythagoras noted that harmonious combinations result primarily from strings with length ratios that are whole numbers. He concluded that music and mathematics must be intricately related in some manner. From this beginning, systems of scales and theories of harmony developed and were studied by mathematicians and musicians alike over many centuries and civilizations right into the modern era.

Among the simplest cases to analyse is the sound generated by a completely isolated, but vibrating, string fixed at its two ends $x = 0$ and $x = L$. By the eighteenth century, the mathematical equation governing the motion of such a string was known. It was a differential equation relating the displacement of the string (which we shall call u) perpendicular to its length, both to the distance x along the string and to time t. For those of you familiar with differential equations its detailed form was (see appendix 2)

$$\partial^2 u/\partial x^2 = (1/v^2)(\partial^2 u/\partial t^2)$$

where v is a velocity determined by the physical characteristics of the string. Since the string is assumed to be fixed at its ends $x = 0$

and $x = L$, the solutions for $u = u(x, t)$ as a function of x and t must obviously include the values $u = 0$ at $x = 0$ and $x = L$ for all t.

The first mathematician to succeed in solving this equation was Daniel Bernoulli (1700–1782) in the year 1775. The Bernoulli family, which in three generations produced no less than eight mathematicians of outstanding ability, are a story unto themselves. Here, however, we are concerned only with one of their second-generation prodigies, Daniel, who became Professor of Mathematics at St Petersburg when he was 25 years old, but settled finally in Basle, Switzerland, where he became Professor of Anatomy and Botany, and finally of Physics.

Using a method known as the 'separation of variables', in which it is demonstrated that a solution can be found as the product of a function of x alone and a function of t alone, he obtained four possible types of solution of the above equation if the 'boundary conditions' (that is the conditions fixing the endpoints $x = 0, L$) are ignored. These solutions are

$$u = \cos(kx) \cos(kvt)$$

$$u = \sin(kx) \cos(kvt)$$

$$u = \cos(kx) \sin(kvt)$$

$$u = \sin(kx) \sin(kvt)$$

where k, for the moment, can be any constant at all with the dimensions of reciprocal length.

Clearly, for our actual string which is fixed at each end, the solutions involving $\cos(kx)$ must be rejected because they do not tend to zero at $x = 0$. We are therefore reduced to possible solutions of the form

$$u = \sin(kx) \cos(kvt)$$

$$u = \sin(kx) \sin(kvt).$$

These are automatically zero at all times at the $x = 0$ end of the string, but are similarly zero at the other end $x = L$ only if $\sin(kL) = 0$. This is true if kL is any non-zero integer (say m) times pi, a condition which now restricts the form of k to

$$k = m\pi/L.$$

The solution with $k = 0$ is of no interest since it reduces to $u = 0$ everywhere (i.e. the string never vibrates at all) and, since the sine of a negative angle is just the negative of the sine of the same positive angle, we can now write a completely general mathematical solution to our vibrating string equation as

$$u(x,\ t) = \sum_{m=1}^{\infty} \sin\ (m\pi x/L)[A_m \cos\ (\theta) + B_m \sin\ (\theta)]$$

in which Σ indicates a sum over m from 1 to infinity, $\theta = m\pi vt/L$, and where A_m and B_m are as yet completely undetermined constants.

Bernoulli was convinced that this infinite sum was the complete general solution for all possible motions of the vibrating string. If so, the parameters A_m and B_m had to be determined by the shape of the string displacements at any time t of interest. This would include the time $t = 0$ (say) of the start of the motion, and hence implies that these constants are ours to choose by the manner in which the string is initially plucked. Now there may not be anything startling to you about this at first glance, but there certainly was to Bernoulli. The problem is that every one of the terms making up the above solution $u(x,\ t)$ is a smooth (we would now say 'analytic') function of distance x along the string, with a well determined slope du/dx at every point x. On the other hand, there is nothing at all to prevent us from starting off the vibrational motion by pulling the string taut at one point (as an archer might do) to create a triangular form $u(x,\ 0)$ with a sharp or pointed apex where the archer's fingers would be (figure 13). Or even worse, we could choose any diabolical starting form which possessed as many 'kinks' as we like. None of these forms is analytical, in the sense of having well determined slopes and curvatures at every point, since the slope du/dx changes discontinuously at 'kink' points. How then can a set of terms, each analytic or smooth, add up to a non-analytic function? The

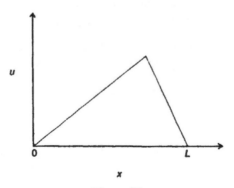

Figure 13

answer, of course, if the proposition is as true as Bernoulli believed it to be, must somehow be connected with the fact that the summation of terms in the solution extends all the way from $m = 1$ to 'infinity'. Unfortunately, in Bernoulli's day there was no technique in mathematics for handling infinite series, so that a rigorous proof that Bernoulli's vibrating string solution was a completely general one could not be accomplished at the time.

Many of the other prominent mathematicians of the day were extremely skeptical. Even the great Leonhard Euler (1707–1783), perhaps the most prolific mathematician of all time, did not think that Bernoulli's solution was sufficiently general. He believed, with many others, that only analytic functions could be represented by trigonometric series. Since the essence of the controversy is present even in the starting condition ($t = 0$) of the Bernoulli solution, we can see it most vividly by putting $t = 0$ into this expression reducing it to

$$u(x) = \sum_{m=1}^{\infty} A_m \sin (m\pi x/L)$$

in which the A_m are constants (i.e. real numbers). The question therefore reduces to whether any function $u(x)$, no matter how smooth or jagged, which is equal to zero at $x = 0$ and $x = L$, can be expressed in this manner for all values of x between the endpoints $x = 0, L$.

The next, and most important, step in our story comes with the work, a generation later, of the French engineer and mathematician Jean Baptiste Joseph Fourier (1768–1830). Fourier arrived at this same mathematical question from a rather different line of research. He had been summoned by the French Minister of the Interior in the summer of 1798 to join Napoleon Bonaparte's expedition to Egypt to promote the advancement of science in the Institut d'Egypte in Cairo, being informed that 'present circumstances have particular need of your talents'. After several military encounters with the British, Napoleon was forced to depart from Egypt in 1801—and Fourier with him. In gratitude for his services, Napoleon appointed Fourier to an academic position in the city of Grenoble, in the French Alps, where Fourier returned to research endeavors. Having suffered in health from the sudden changes of climate to which he had been subjected, his primary research efforts were now concentrated, perhaps not wholly surprisingly, on the study of heat—and in particular on heat loss by radiation, and heat conduction in solids. With respect to the last problem, Fourier found that the differential equations

Baron Jean Baptiste Joseph Fourier, 1768–1830. Sketch of Fourier as a young man by his friend Claude Gautherot. Reproduced by permission of AIP Emilio Segrè Visual Archives. *Physics Today* collection.

involved had many features in common with that of the vibrating string. Like Bernoulli before him, he also derived solutions as infinite sums of trigonometric functions, although it was clear from the physics that the answers must relate as much to discontinuous functions as to continuous ones. The results were first presented at a memorable session of the French Academy on 21 December 1807. It was in this session that Fourier announced his thesis that *any* arbitrary function, defined in a finite interval by a graph of any kind whatsoever, can always be resolved into a sum of pure sine and cosine functions.

This claim went considerably beyond that of Bernoulli by removing a number of restrictions contained in the earlier problem. In particular, in the context of the vibrating string, Fourier realized that the range of x in that problem (namely, $0 < x < L$) did not coincide with the natural period of the fundamental sine function $\sin(\pi x/L)$ involved in the solution, which was $2L$ rather than L. However, if the solution for $u(x)$ was allowed to extend

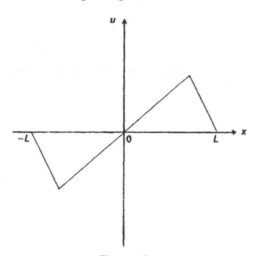

Figure 14

also in the negative x-direction, then this impediment could be removed by defining a function now running between $x = -L$ and $x = L$ for which $u(-x)$ is just the negative of $u(x)$. Such a function is referred to as an 'odd' (as opposed to 'even') function, and an example, based on the archer's bow curve of figure 13, is shown in figure 14.

As an indication of the manner in which an infinite series of sine functions can converge to a non-analytic resultant we sketch, in figure 15(a), the first few terms for the 'rectangular' curve shown in that figure. There are, of course, well defined methods (involving integral calculus) of finding the exact values for the constants A_m involved in the series although such details need not bother us here. For the rectangular function of figure 15 the actual series begins

$$u(x) = A[\sin (y) + (\tfrac{1}{3}) \sin (3y) + (\tfrac{1}{5}) \sin (5y) + (\tfrac{1}{7}) \sin (7y) + \ldots]$$

where $y = \pi x/L$, and the series continues *ad infinitum* in the pattern made obvious by the first few terms. The constant A determines the 'height' H of the rectangle according to the formula $H = \pi A/4$, and the manner in which the series gradually approaches the limiting rectangular form can now be imagined from the curves labeled 1, 2, 3, 4 in figure 15(a), which show respectively the above series terminated at the first, second, third and fourth terms. Although the convergence is slow, those non-analytic square corners begin to be approximated better and

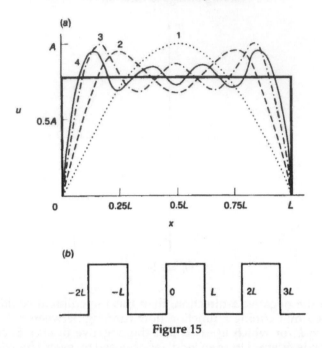

Figure 15

better (i.e. with less and less roundedness) as progressively more terms are included. It certainly seems quite suggestive that, in the infinite limit, the representation would be exact. And so it seemed to Fourier although, like Bernoulli before him, he still possessed no methods for rigorously dealing with extrapolations to infinity. If the sine terms are extended beyond the range $0 < x < L$ (of figure 15(a)) in both the positive and negative directions, the limiting curve of figure 15(b) is obtained which demonstrates the 'odd' character $u(-x) = -u(x)$ of the sine representation.

Fourier now reasoned that, if the infinite limit of the above sine series truly did reproduce the non-analytic rectangular form of figure 15(a) between $x = 0$ and $x = L$, and $u(x)$ did remain precisely equal to H as x approached arbitrarily close to zero, then the argument used to exclude cosine terms from consideration (i.e. they were all non-zero at $x = 0$) was no longer relevant. In fact, he showed that the very same rectangular function in the range $0 < x < L$ could just as well be represented by a cosine series in the form

$$u(x) = \sum_{m=0}^{\infty} B_m \cos(m\pi x/L).$$

However, unlike the sine series, the cosine one describes an 'even' function which, when extended into the negative x-domain obeys the relationship $u(-x) = u(x)$. With a little more thought it now became clear to Fourier that, since any function whatsoever defined in a range between $x = -L$ and $x = L$ can be expressed as the sum of an 'even' and an 'odd' function pair in the manner

$$u(x) = \tfrac{1}{2}(u_{\text{even}} + u_{\text{odd}})$$

where

$$u_{\text{even}} = u(x) + u(-x)$$

and

$$u_{\text{odd}} = u(x) - u(-x)$$

then any such function must also be expressible as an infinite series containing both sines and cosines. We show such an odd and even function decomposition for an extremely 'arbitrary' function in figure 16. Its 'Fourier' representation in terms of sines

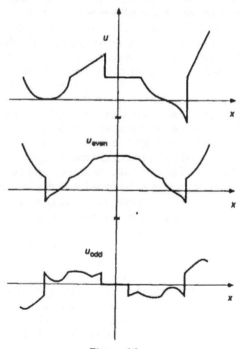

Figure 16

and cosines will, of course, have a periodicity of $2L$, but since L can be made arbitrarily large, Fourier's claim amounted to the statement that any function whatsoever, whether smooth or jagged, finite or infinite, can be expressed as a series of sines and cosines.

Unfortunately, when Fourier first made public this somewhat startling pronouncement to the French Academy on that December day in 1807, his work was by no means universally accepted. The lack of rigor in its discussion of the infinite limit was the point of contention, and the eminent trio of French mathematical moguls of the day, Laplace, Lagrange, and Legendre, all focused on this weakness and expressed their reservations in no uncertain terms in a written report stating that 'the manner in which the author arrives at his equations leaves something to be desired in the realms of both generality and rigor'. Fourier protested, but to no avail, and his work was not published by the Academy. But he was ultimately to prevail and, in 1822, he gathered all his work together and published it in a monumental treatise entitled *The Analytic Theory of Heat*, which was undoubtedly one of the most influential works of mathematical physics to appear in the nineteenth century.

Fourier's methods enabled him to work with discontinuous functions when others were still wrestling with the continuous, and to discuss concepts like 'convergence' before there was even an accepted definition of such an abstraction. He even introduced 'functions' that were infinite at one point and zero everywhere else. And the justification, in the absence of mathematical rigor, lay in the success of his work in applications to subjects as diverse as heat, hydrodynamics, acoustics, and electrodynamics. As a result of these successes, a new generation of mathematicians felt encouraged to try to build a more rigorous mathematical foundation beneath Fourier's utilitarian concepts.

The immediate problem to be faced was the fact that there was, at the time, no precise definition of convergence. The first effort to introduce some rigor in this context was made by the French mathematician Augustin-Louis Cauchy (1789–1857), although he concluded (incorrectly as we now know) that the sum of a convergent series of continuous functions must also be continuous. Major additional refinements, leading to a correction of Cauchy's statement, were made by a young University of Berlin professor (who was later to succeed Gauss at the University of Gottingen) named Peter Gustav Lejeune Dirichlet (1805–1859) and, in particular, by one of his students Bernhard Riemann, whom we met in a different context in the last chapter. However, such were the complexities of the infinite that almost a century would pass from Fourier's initial pronouncements in 1807 before all the details of a fully rigorous proof of his ideas were finally achieved. Included in the final analysis was the recognition that Fourier's 'theorem' was not, in fact, quite as all-embracing as he had thought. Certain limitations had to be placed upon the curves in question. In particular, it was recognized that some curves can be defined which are so pathological in their behavior that even an infinite sum of sines and cosines is not sufficient to reproduce them—for example, a curve must not be discontinuous, or more exactly not differentiable, at *every* point (i.e. a 'fractal' in modern parlance; see chapter 11). But even though the precise restriction is a little more liberal than this, curves which cannot be resolved into 'Fourier components' are still relatively uncommon in the mathematical physics literature.

It is interesting to note that the lack of rigor of Fourier's theorem during the nineteenth century, and the countless attacks upon it from the standpoint of pure mathematics, in no way prevented it being put to good use during this period by applied mathematicians and physicists. Once again, this is an interesting contrast of the approaches of the different disciplines. Pure mathematicians, armed with the weapon of sharp and rigid proof, tend to have little use for any alleged theorem until it can successfully withstand the severest criticism of the day. Scientists, on the other hand, are interested primarily in the interpretation of experiment, and are fully aware that absolute accuracy is never attainable in their field. For them, therefore, any mathematical 'tool' that appears to assist in experimental interpretation is happily employed, and is discarded only if found to be wanting at the level of the experimental precision available.

In the nineteenth century, the Fourier theorem was particularly useful in the context of what are known as boundary problems. These are motional problems subject to equations much like that given at the beginning of this chapter with reference to the vibrating string, but containing much more complicated 'boundary' restrictions. For example, the vibrations of a drum might fall into this category, since the shape of the drum periphery (along which line the drum membrane is constrained not to vibrate at all) is a relatively complex boundary condition. The motional solutions for the Fourier decomposition of membrane vibrations (and therefore the forms of the sound vibrations emanating from the drum) obviously are a sensitive

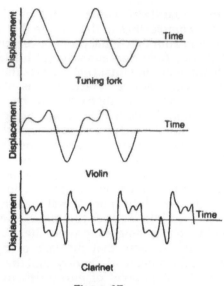

Figure 17

function of drum shape. In fact, the resolution (that is, decomposition) of the sounds of various musical instruments into their Fourier components was an active topic of acoustical physics in the mid-1800s. One of the prime contributors was the German scientist and philosopher Hermann Helmholtz (1821–1894) who authored a number of books on acoustical and optical perception in the 1870s.

The simplest musical tone of all is that of the tuning fork, which is close to a single sinusoidal wave. The composition of the tone of the violin is a little more complex and that of the clarinet more complicated still (see figure 17), and so on down the line through trumpet, saxophone, etc all the way to the above-mentioned drum. Along this progression, the number of Fourier components required to analyse the measured 'shape' of the vibrational spectrum gradually increases, but none of the shapes ever seems to become too 'jagged' to be described uniquely in Fourier terms.

In spite of the usefulness of Fourier's theorem in the context of classical vibrational problems, the real explosion of activity making use of this most valuable of mathematical tools came in the twentieth century with the realization that the atoms in crystals are arranged in periodic patterns. This fact, which was first experimentally demonstrated by x-ray diffraction just before the First World War, means that every crystal possesses a smallest volume (called its 'primitive unit cell') from which the entire crystal can be built up, at least in principle, by simple repetition (or stacking together). The simplest of all such systems would be one composed of a single atom-type in a simple cubic arrangement, for which the primitive cell is just a cube with atoms at the corners. For this crystal all atomic sites can be defined in terms of three integers (say i, j, k) which measure their distances along the three coordinate directions x, y, z in units of the side length 'a' of the primitive cell. For more complicated crystals the primitive unit cell may be of a more elaborate shape, and may contain many atoms of different types inside it. But even if this is so, the numbers i, j, k still play the essential (if no longer the sole) role in defining exactly where the atoms are. Most importantly, they are responsible for describing the periodicity of the lattice.

To make the mathematics involved as simple as possible, we shall first imagine a one-dimensional 'crystal' containing just a single atom-type. This 'crystal' is then simply a linear chain composed of an extremely large number N of identical atoms, each a distance 'a' from its neighbor. Mathematically we can therefore define the general equilibrium position x_n of the nth atom along the chain (which we consider to extend in the x-direction) in the manner $x_n = na$. The integer n can run from 1 to N and in crystals of macroscopic length (say a centimeter or two) the number N is of order one hundred million, or 10^8. Physically, these atoms are held in place by interatomic forces which, however, do not restrict them rigidly, but allow them to vibrate in an almost sinusoidal (or what physicists call 'harmonic') fashion about their mean positions. As a simple picture (see figure 18(a)) each atom, of mass M, can be thought of as attached to its nearest neighbor on either side by an invisible spring. In this situation, the equation of motion for the x-axis displacement u_n of the atom on atomic site n from its mean (or equilibrium) position $x_n = na$ can be shown to take the form

$$M(\partial^2 u_n/\partial t^2) = g(u_{n-1} - u_n) - g(u_n - u_{n-1}).$$

Now this equation is not really as mysterious as it might look at first sight because it has a simple physical interpretation. Thus, the left-hand side is just mass times acceleration for the nth atom. By Newton's laws of motion it should therefore be equal to the resultant force acting upon it. The latter is the difference between two opposing forces, one pulling it to the 'left' and the other to the 'right'. These are the two terms, each multiplied by a 'force

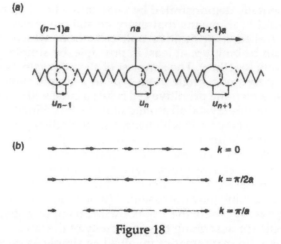

Figure 18

constant' g, seen on the right-hand side of the equation with a negative sign between them. The fact that the individual forces are each linear in the displacement variable u is a reflection of another famous law of physics known as Hooke's Law.

But enough about physics! We are really interested in the mathematical form of the equation and in how we are going to solve it. It differs from the equation of motion for a string, with which we started this chapter, in one essential way. Thus, whereas the second differential of the displacement u at the point x of the string problem depended only upon the value for u at that very same point, in the equation for an atomic vibration this same differential at one value for x (e.g. $x = x_n = na$) depends not only upon the value for u at this same site, but upon the values for u at the nearest-neighbor sites x_{n+1} and x_{n-1} as well. We say that the atomic equation of motion is not 'separable' in the variable u_n, in the sense that we cannot write for it a form which contains u_n alone. What we need to find is a new variable, related to the various u-values at the different sites, whose equation of motion contains 'only itself'. And it is here where the Fourier series comes to the rescue.

Writing a sine or cosine Fourier series as follows:

$$u(k, t) = \sum_{n=1}^{N} u_n \cos (nka - \omega_k t)$$

we define the function $u(k,t)$, which is called the 'Fourier transform' of the displacements u_n. From its definition, it represents a *traveling* wave motion of frequency ω_k in which all N of the atoms in the chain participate. We shall see in a moment that our equation of motion can be written entirely in terms of $u(k, t)$, so that it is this function that exists as an independent entity in the atomic vibrational problem. First, however, we ought to give a little thought to the quantities entering this Fourier transform. What, for example, is this parameter k which, together with the time t, labels the independent excitations, and what values can it take? Obviously it is closely related to its namesake in the vibrating string problem and, consequently, its allowed values depend on the manner in which we constrain the motion of the atoms at the end of the chain. Equally obviously, however, if the chain is some 10^8 atoms long, the consequences of this decision cannot have any great impact on the physics of the situation. The usual assumption is to imagine the chain of atoms joined upon itself in such a way that u_{N+1} is identical with u_1, u_{N+2} with u_2, and so on. If this is the case, the above Fourier transform could just as well have been defined with the sum on the right-hand side running between $N+1$ and $2N$ as between 1 and N. The two forms must be identical, which tells us that

$$\cos (nka) = \cos [(N + n)ka].$$

This, in turn, implies that Nka must be some integer (say m) times two pi or, in symbols,

$$k = 2\pi m/Na.$$

Physically different solutions are obtained when m assumes any of N consecutive integer values. The usual convention is to adopt a symmetric situation with m running between $-N/2$ and $N/2$. The quantity k can then run between $-\pi/a$ and $+\pi/a$.

We have gone into all this detail only because this Fourier-generated k-quantity is arguably the most important concept in all of solid state physics. In the linear chain case it is called a 'wave number' but, in real three-dimensional problems it becomes a vector quantity and is referred to as a 'wave vector'. Since the quantity 'a' is a length (specifically the side length of the primitive unit cell) it follows from the above equation that k has the dimensions of reciprocal (or inverse) length. It is said to exist in 'reciprocal space', which is just like any other space except that within it the distance between points is measured in cm^{-1} rather than in cm. For the linear chain problem, the 'reciprocal lattice', defined by the allowed values for k, is just a dense set of points

for transmitting sound in solids. With wavelengths that are enormously longer than the distance 'a' between the atoms, they are not particularly sensitive to the discreteness or periodicity of the lattice. They are called 'acoustic waves' (although most of them have frequencies higher than can be perceived by the ear) and can exist quite happily in liquids or even gases. For example, the acoustic (or pressure) waves in air produced by the notes in the middle of a piano have wavelengths of about 3 or 4 feet. These waves, in figure 19(a), would appear less than one ten billionth of the way out to the zone edge at $k = \pi/a$. The slope of the acoustic mode curve near the Brillouin zone center is the acoustic velocity and, as can be seen from the figure, it is essentially independent of wavelength so long as the wavelengths are large compared with atomic dimensions. In crystals this velocity is typically a few miles per second while in air it is about a factor of ten smaller; we call this the 'speed of sound'.

The shortest wavelength lattice vibrations present in the single-atom-type chain have a wavelength equal to $2a$ (see the pattern in figure 18(b) with $k = \pi/a$). In crystals, these vibrations are of extremely high frequency and, since they are sensitive to the periodicity of the chain, they are not so well defined in liquids and gases. In the two-atom-type chain, modes of this same character (with nearest-neighbor atoms moving out of phase) have been transformed to the center ($k = 0$) of the upper spectral branch of figure 19(b). If the two-atom-types involved have different electronic charges (like positively charged sodium and negatively charged chlorine in table salt) then modes of this kind couple very strongly to electric fields, such as those carried by light waves. As a result, the upper curve of figure 19(b) is called the 'optic mode' branch of the vibrational spectrum, with the lower curve then being the 'acoustic branch'.

Real crystals, of course, are periodic in all three dimensions, and nearly always have at least two kinds of atomic constituent. The mathematics of the Fourier transform now naturally becomes more difficult and often requires computer calculation, but the principle remains the same. The independent modes of atomic vibration are still defined by a quantity k, which is now a three-dimensional vector defined inside a three-dimensional Brillouin zone. At each value for k there will be $3n$ modes of the vibrational spectrum, where n is the number of different atoms in the primitive unit cell of the crystal. The shape of the Brillouin zone will depend upon the particular symmetry of the crystal structure, but all the modes so defined will still be simple traveling sinusoidal waves. The lowest-frequency branches (now three of them, one for each of the three dimensions) with frequencies tending to zero at the Brillouin zone center still carry the sound waves. The rest are various types of 'optic mode' vibration. In most cases, the three acoustic modes at any particular k-value can be separated into one which vibrates along the direction of propagation of the wave (called the longitudinal mode) and the other two which vibrate at right angles to each other and to the direction of wave travel.

The Fourier transform method in solid state physics depends on the existence of a periodicity in the crystal structure. The immense value of this technique is brought home to physicists when they confront similar problems in glasses, for which the periodicity is not there. For these cases the equations of motion remain in the form of some 10^{24} variables all intertwined. Independent modes of vibration must still exist, but it is virtually impossible (even with the most powerful computers) to describe them with any precision. The physics of glasses therefore lags far behind its crystalline counterpart. In crystals the Fourier transform is 'king'. The lattice vibrational problem can be solved (albeit on occasion, with the help of numerical methods) and, as a result, the independent modes can be defined. With the coming of quantum theory (see chapter 12) each of these modes could be treated separately as a quantum oscillator, acoustic and optic vibrational modes becoming acoustic and optic 'phonons' in the new language. In this manner the entire field of low-temperature thermal physics in crystals was opened up to quantitative analysis. The man primarily responsible for recognizing the value of the Fourier method in the lattice vibrational context was the Dutch–American theoretical physicist Peter Debye (1884–1966) and the field is now commonly named after him—the Debye theory of solids.

In more recent years the applications of these quantized modes of lattice vibration have been extended far beyond anything that Debye could have envisaged. Most important was the recognition that these 'phonon' modes are only truly independent when the amplitude of the oscillations that they (classically) describe is infinitesimally small. Only in this limit are the equations of motion truly linear in the displacements u (Hooke's Law) and the solutions exactly sinusoidal or 'harmonic'. In real crystals at room temperature the oscillations are large enough that interactions between the phonons become important. In particular, these interactions can induce a temperature dependence on the frequencies ω_k and therefore also on the curves like those in figure 19. A whole branch of theoretical physics involv-

Peter Debye, 1884–1966, Nobel Prize for Physics 1936. Reproduced by permission of Mary Evans Picture Library.

ing changes of crystal structure with temperature has grown out
of this effect. If, as a function of temperature, one of the modes of
wave vector k is 'driven' to zero frequency by the phonon
interactions, then the symmetry of the crystal structure changes
at this temperature. Referred to as the 'theory of soft modes', an
enormous modern literature has developed couched in these
terms and incorporating such phenomena as ferroelectricity (or
electrically 'charged' crystals) and ferroelasticity (or spontaneously 'strained' crystals).

Even more widely, *any* property associated with the individual
atoms in crystals must have this same structural periodicity
which enables the Fourier method to be applied. Another
example of immense importance in modern-day solid state physics concerns the elementary (that is 'atomic') magnets associated
with many kinds of atoms. These tiny atomic magnets, one on
each magnetic atom, also interact with each other to produce
magnetic oscillations. As before, the independent modes can be
found by using Fourier transforms, and they are labeled once

more by the all-important wave vector k. In quantum language these independent magnetic oscillations are called 'spin-waves' or 'magnons'. They travel, just like lattice vibrational waves, throughout the crystal and possess their own ω_k relationships, velocities, and Brillouin zones. It is through them that all the modern theories of magnetism in crystals have been developed.

There are, however, even more difficult and important problems connected with the properties of crystals about which periodicity and Fourier transforms have all-important contributions to make. These problems concern the motion of electrons. In many crystals, electrons do not stay localized in the vicinity of their 'parent' atoms, but can run fairly freely throughout the whole crystal. Although some of the best electronically insulating solids do have electrons that are bound to their parent atoms (and therefore this lack of electron mobility immediately explains the absence of electronic current flow in the presence of an applied voltage) many other solids, both insulators as well as semiconductors and metals, all contain electrons that are comparatively free to wander throughout the whole sample. For these, the reason why some of them conduct electricity while others do not is not at all obvious and is, in fact, at least at first glance quite baffling.

In the context of electronic motion the decisive breakthrough came in 1926 with the creation of 'wave mechanics' by the Austrian physicist Erwin Schrödinger (1887–1961), whom we shall meet again in chapter 12. In wave mechanics the electron is defined completely by a spatial function $\phi(x, t)$, called the wave function, the square of the magnitude of which gives the probability of the electron being at the particular point x at the time of interest t. Although it is not necessary, in the present chapter, to pursue any of the details of Schrödinger's 'wave equation' for the motion of $\phi(x, t)$, we can say that it is a differential equation which contains within it a function describing the interactions between the electron in question and the environment in which it finds itself. These interactions are primarily with the atoms of the crystal and they therefore possess the periodicity of the atomic lattice in question. Because of this periodicity, the Fourier transform again becomes the essential tool for solving the wave equation for the motion of an electron in a crystal.

The very simplest situation of all, of course, would concern the motion of a completely free electron in space; that is, subject to no interactions whatsoever. For this case the Schrödinger wave equation has a traveling wave solution of the form

$$\sin (kx - \omega t)$$

where the wave vector k and frequency ω are related respectively to the momentum p (i.e. mass times velocity) and energy E of the electron in question by the equations

$$2\pi p = hk \qquad 2\pi E = h\omega$$

where h is an universal constant of nature (called Planck's constant) with a value that can be directly determined by comparison of quantum theory with experiment. Thus, for a free electron, k can take on a completely continuous and unrestricted set of values determined solely by the electron momentum. Also, since the energy and momentum of a particle are known in physics to be related by the equation $E = p^2/2m$, where m is the particle mass, it follows from the above equations that E is proportional to k^2 for the free electron. We show this relationship in figure 20.

If we now consider the electron to be traveling inside a crystal and interacting with the atoms, the periodic array of the atoms makes the interaction forces periodic. If we imagine, again for mathematical simplicity, a linear chain crystal, the resulting Schrödinger equation can be solved by transforming to reciprocal space and defining a Brillouin zone exactly as was done for the atomic vibrational problem. The solution consists again of traveling waves, but now of the slightly more complex form

$$u_k(x) \sin (kx - \omega t)$$

with $u_k(x)$ having the periodicity of the lattice and details which depend on the mathematical form of the interaction forces between the electron and the atoms. The most important feature of the solutions, however, is the manner in which they perturb the simple-free-electron continuous relationship between k and energy E shown dashed in figure 20. The new relationship is as shown by the full curves in figure 20. The electron–atom interactions have opened up gaps in the 'energy spectrum' at the Brillouin zone boundaries. That is to say, some energies now exist for which no traveling electron solutions can be found. Since the wave vectors outside the first Brillouin zone can always be directly related to those inside the first zone, it is more common to see these energy 'dispersion relationships' with all the curves 'folded back' inside the first zone (between $-\pi/a$ and $+\pi/a$), which then exhibits a series of vertically stacked 'energy bands' separated by 'energy gaps'.

In a real crystal, of course, not only will the Brillouin zone be three-dimensional, but there will be far more than a single electron present. There will, in general, be an extremely large

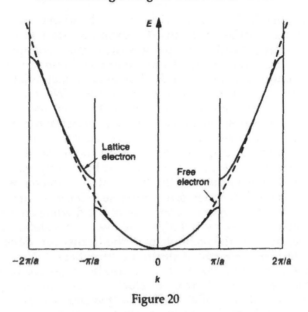

Figure 20

number of traveling wave electrons in the system; say n for each of the N atoms in the crystal. For electrons (unlike phonons) the laws of quantum physics dictate that there can be no more than two electrons occupying any one particular wave vector state. Consequently, as a general rule, the electrons present will fill the energy 'bands' of figure 20 from the bottom up. Since there are N wave vector states in the first Brillouin zone, the final picture is one of completely filled and empty energy bands if n is an even number, and of the presence of a half-filled band if n is an odd number.

In any one filled band, since each $+k$ wave vector has a $-k$ counterpart, there are as many electron waves traveling in any one direction as in its opposite, so that no resultant current flows. When an electric voltage is applied, electrons can be rearranged, but only by energy amounts corresponding to the magnitude of this voltage. Since practical voltages can only produce energy shifts which are small on the scale of the energy gaps, it follows that only systems with partially filled energy bands can respond to exhibit electrical conductivity. We call them metals. Filled band systems are all electronic insulators, in spite of the fact that they contain traveling electron waves running throughout the crystal.

Naturally there are other complexities in real systems that we have thus far passed over. Real crystals, with their three-dimensional Brillouin zones, can exhibit band overlaps at different points in different bands (for example, state $k = k_1$ of the 'upper' energy band may dip below the highest energy state, say $k = k_2$ of the 'lower' energy band). In this way one can sometimes obtain band overlaps (which, however, affect relatively few electrons) even for cases when n is even. We refer to these situations as examples of 'semi-metals'. In other cases, band gaps may exist but are smaller than thermal energies at room temperature. For these materials the properties vary in a unique fashion as a function of temperature, being insulators at low temperatures and relatively good conductors at high temperatures; they are called semiconductors. By doping these systems with small amounts of other elements, it is possible to induce extra electrons to appear in the upper or 'conduction' band while creating extra electron vacancies, or 'holes', in the lower or 'valence' band. In this manner was the technology of the 'transistor' developed to revolutionize the world of electronic amplifiers.

Another complication which has also been glossed over to this point is the fact that electrons interact with each other to produce their own relative oscillations (called 'plasmons' in the language of quantum theory). In addition, electrons also interact with the lattice vibrations, or phonons. This interaction can cause the electrons to behave as though they are dragging lattice distortions with them, and we refer to the resulting 'dressed' electron as a 'polaron'. Electron–phonon interactions can also, in some materials at very low temperatures, produce an effective binding together of $(+k, -k)$ pairs of electrons in a fashion that produces superconductivity—or a complete vanishing of all opposition whatsoever to electric charge flow. Phonons themselves can also sometimes interact directly with light to produce a mixed light-wave–lattice-wave oscillation called a 'polariton'. The combinations seem almost endless. But the important point, from the perspective of this chapter, is that all these strange phenomena in crystalline solids are understood in the language of the wave vector. Life in modern theoretical solid state physics is conducted almost entirely in reciprocal space. No solid state physics text book can any longer be written without the Fourier transform appearing very early in the text and permeating the entire volume. In fact, so commonplace has life in reciprocal space become to the solid state physicist that the occasional 'event' requiring reversion to 'real' space for its understanding (some impurity phenomena, for example) strikes him or her as something of an aberration. Everything moving in a crystal seems to live in k-space, interact with other k-space entities, and (if necessary) decay and die in k-space.

The true realization of the immense reliance of solid state physics on the Fourier transform only really struck home in the 1970s when theoretical physicists made their first serious efforts to move on from understanding crystals (with their nice periodic atomic structures) to understanding liquids and glasses, which lack this feature. In these materials the k-vector cannot be defined precisely, and the comfortable environment of reciprocal space becomes badly mauled. To a large degree science is then forced back into real space—and the vision is not a pleasant one. Most progress, to date, has come from computer modeling using giant 'number-crunchers'. Most structural excitations just do not propagate in simple sinusoidal forms over long distances, and they no longer interact in a nice mathematically simple manner with other excitations as they did in crystals. Everything seems to be badly intertangled, excitation lifetimes are small, interactions are complicated. What is missing is a mathematical 'key' to untangle this mess of equations; one to perform for liquids and glasses what the Fourier transform did for crystals. Is it, perhaps, lying there unused in the archives of pure mathematics? Is there another giant upon whose shoulders a better theory of non-crystalline phenomena can be built? Most physicists doubt it—but deep down, there is a feeling that nature should surely not be quite as unfathomably complicated as our present efforts to understand the 'amorphous' state imply; so that there is always hope!

5

The Perception of Number: from Integers to Quaternions

In the beginning numbers were for counting (possibly sheep at bedtime). But over the centuries, the concept gradually developed from the simple starting point of the counting numbers 1, 2, 3, 4, ..., first to include zero, and then to embrace the idea of negative integers −1, −2, −3, −4, ... The latter are, in some sense, not quite so obviously 'real' as the counting numbers (or positive integers, if you prefer) since, for example, no-one has ever seen minus two cows or minus two dogs. Indeed, to the Ancient Greeks, the equation $x + 2 = 0$ had no solution. However, negative numbers are essential (from a formal standpoint) if the concept of subtraction is to be made perfectly general. In other words, only with negative numbers available can you *always* perform subtraction. Without them you can do some subtractions, like

$$8 - 3 = 5$$

but not others, such as

$$3 - 8 = ?$$

With negative numbers available the latter subtraction is no longer a problem. Using them you can now write

$$3 - 8 = -5$$

and even learn to give them an interpretation in real life. For example, that −5 above might well be the worldly representation of the five dollars you still need to add to the three in your pocket before that eight dollar purchase can be made. In spite of this, the step from positive numbers to negative numbers was historically not an easy one to take, and they were not fully incorporated into mathematics until well into the sixteenth century, although there are some isolated examples of their appearance (primarily in India) much earlier. The systematized arithmetic of negative numbers (and of zero) is first found in the work of the Hindu mathematician Brahmagupta, who lived in central India in the seventh century.

In spite of the conceptual difficulties associated with negative numbers, the basic rule for multiplying two negatives, such as in the equation

$$(a - b)(c - d) = ac + bd - ad - bc$$

was perfectly well known in Ancient Greek times. They would accept this result, however, only if the two numbers to be multiplied, that is $(a - b)$ and $(c - d)$, were both positive. For example

$$(5 - 2)(6 - 3) = 30 + 6 - 15 - 12 = 9$$

was perfectly acceptable, whereas

$$(5 - 2)(3 - 6) = 15 + 12 - 30 - 6 = -9$$

was not. Moreover, although the Greeks had a concept of nothingness, they never interpreted it as a number. Brahmagupta, to his credit, did; but he also blotted his copy-book by asserting that zero divided by zero was zero and, on the touchy subject of 1, 2, or 3 divided by zero he did not commit himself.

With the set of integers properly expanded to include both positive and negative numbers (and zero) we can always add and subtract, but we cannot always divide. Some divisions present no problem (such as $6/2 = 3$); but what about dividing 6 by 7? This division obviously has no answer within the family of integers so that, in order to make division perfectly general, it is necessary to enlarge the number system once more. What is required is the notion of fractions.

Generally, the problem of dividing any arbitrary integer n by any other m (positive or negative) is 'solved' simply by defining new numbers to be written as n/m. These are the 'rational' fractions, implying that they are the 'ratio' of two integers. A particular convenience of these rational fractions is the fact that they are not new numbers in the sense of being completely separate from the integers, since they include the latter as special cases. Thus, for example, $3/1 = 3$ and $5/-1 = -5$. With this extension of the concept of number (called the 'rational number' system) we can now always add, subtract, multiply and divide without running into any problems. Well almost; one difficulty that does remain concerns dividing by zero. This operation cannot be consistently defined within the existing framework since it, alone, does not lead to another member of the rational number set which has thus far been constructed. We therefore take the easy way out and simply exclude a division of this kind as 'not allowed within the rules of the game'.

There are, of course, an infinite number of fractions, just as there are an infinite number of integers. But the new extended notion of the meaning of 'number' is very different from the previous set of counting (or 'natural') numbers. For example, if we consider a general fraction n/m in which n is not zero (and m is not allowed to be zero) then, no matter how small a value this fraction may have, we can always divide it by 2, and by 2 again, and again, and again, . . ., and so on forever. This means that we can always create a limitless number of fractions between any given small fraction and zero. It follows that the rational numbers are very 'dense', or 'close together', in this sense. By defining them we have, in a way, 'filled in the gaps between the integers' and it is tempting (although, as we shall soon see, incorrect) to think that we have filled them in completely. On the other hand, by using the rational numbers we can certainly get as close as we please to any number of interest.

The rational numbers therefore seem to be just the tools necessary for measuring and, for the cases where n and m in the fraction n/m are both positive integers, they have probably been used in this capacity since the Bronze Age. Fractions with unit numerators (i.e. $\frac{1}{2}$, $\frac{1}{3}$, $\frac{1}{4}$, etc) often appear on Egyptian papyri (disguised, naturally, by their Egyptian hieroglyphic notation). On the other hand, fractions with other than one in the numerator tended to be avoided in explicit notation in the Egyptian world and were usually re-expressed as in the form

$$\tfrac{2}{3} = \tfrac{1}{2} + \tfrac{1}{6}.$$

The complete use of the entire set of positive rational fractions appears to have been used by the Babylonians perhaps as long as 4000 years ago. Their emphasis was on adaptation as a practical system of spatial measurement, and emphasis on the 'ratio' aspect of fractions did not really appear much before the Euclidean era of Greek mathematics. This view tended to sharpen the theoretical aspects of the concept of number, accentuating the connection between pairs of integers and downplaying their role simply as a tool of measurement.

The first real bombshell in the development of numbers took place about 500 BC (give or take a few decades) and is often attributed, but very likely erroneously, to Pythagoras (582?–502? BC). To that time is had been a fundamental tenet of mathematics, particularly as it relates to geometry, that all measure is explainable in terms of the positive integers and their ratios. At this time, however, the Greek mathematical community was stunned by the demonstration that, lurking within the simplest of geometric constructions, were 'numbers' that could not be found among the rational fractions.

Consider, for example, the problem of constructing a square with an area exactly double that of the unit square (or, in other words, construct a square with an area of two square units). To find the length of its sides, we seek the rational number n/m whose square is equal to two. In symbols this becomes $(n/m)^2 = 2$ or, rearranging the terms a bit

$$n^2 = 2m^2.$$

If the fraction n/m is expressed in its simplest form by dividing the top and bottom by any common factors that exist, it follows that n and m cannot both be even (otherwise we could divide both the top and bottom by two). Looking at the above equation we see that the square of n is certainly even, since it is two times an integer. Moreover, since only even integers have even squares, n itself must be even and, therefore, m must be odd.

This much decided we now write the even number n as two times some other integer, say $n = 2k$, and substitute it into our defining equation $n^2 = 2m^2$ to derive the relationship $4k^2 = 2m^2$ or, equivalently, $2k^2 = m^2$. This tells us that the square of m must be even and hence (again since only even numbers can have even squares) m itself must be even. The rational fraction n/m sought is therefore required to have m both even and odd. Since this cannot be, the associated fraction n/m does not exist. The conclusion is that, densely spaced though they are, rational numbers are just not dense enough to include the number which, when squared, is equal to two. We refer to this number as the square root of two, and write it $\sqrt{2}$, with a fancy symbol, simply because we have no other concise way of writing it. Note, in particular, that (via the Pythagorean theorem) this $\sqrt{2}$ is the length of the diagonal of the unit square, so that its embarrassing presence in elementary geometry could hardly be more glaring. And there is nothing special about the choice of the number 2 in all of this. It is just as easy to prove that all square roots $\sqrt{n}$, for which n itself is not a perfect square, are equally inexpressible as rational fractions. They all fall into some new category which, perhaps not too surprisingly, is now called the set of 'irrational' numbers.

Much as the notion of irrational numbers tormented the Ancient Greeks, they surely would have been even more concerned had they known (as we now know) that there are infinitely more irrationals than there are rational numbers. Although a rigorous demonstration of this, and even a proper understanding of what it means, requires an ability to distinguish between different infinite quantities (something that was not acquired until the 1870s) some inkling that this may well be the case can be gleaned by expressing rational fractions as decimals. The decimal system, as we know it (complete with the notation of the decimal point), was developed towards the end of the sixteenth century in France and, using it, it is not difficult to prove that all rational fractions have a decimal expansion in which the digits finally 'cycle', or repeat themselves, endlessly in a simple pattern. Thus, for example,

$$3/8 = 0.375\,000\,000\,00 \ldots$$

$$2/11 = 0.181\,818\,181\,818\,1818 \ldots$$

$$3/7 = 0.428\,571\,428\,571\,428\,571 \ldots.$$

Although we do not usually write down the endless string of zeroes in the first example above, they nevertheless express what is perhaps the simplest of all forms of 'cycling'. In other examples, the cycling pattern may not manifest itself right away, but appear eventually it must, and with a number of digits in the repeat cycle no longer than one less than the denominator of the fraction.

Since all rational numbers are cycling decimals (and conversely all cycling decimals are rational numbers) it follows that the irrationals must be made up of an endless string of decimals which never cycle. The rational numbers therefore possess a very special property and it is not difficult to appreciate that there are countless different ways of interfering with any particular cycling decimal to destroy this property. Thus, from the decimal point of view, the rational numbers are only a tiny subset of all numbers.

Armed with the rational and irrational numbers (together referred to as the 'real' numbers) mathematicians became able to express exactly the position of any point on a 'number line', at least in principle. Unfortunately, this number system was still not sufficient to solve all the mathematical problems that came their way. For example, by using them they could always add, subtract, multiply and divide (properties which are said to define a 'number field') but they couldn't always find powers and roots. This means that there are still elementary algebraic equations which cannot be solved. For example, at an extremely elementary level, we still cannot solve the equation $x^2 + 1 = 0$. Using the familiar symbolism, the number we seek is $x = \sqrt{-1}$. But no rational or irrational number (positive or negative) will, when squared, give -1 and hence solve the equation. The required number, whatever it is, was still outside the number system as it existed up to the dawn of the sixteenth century. In fact, well into the seventeenth century and, in many cases, even beyond that, numbers like the square root of -1 were treated as objects of supreme mathematical mysticism. The French philosopher and mathematician René Descartes insisted on referring to them as 'imaginary', while the German mathematician Gottfried von Leibniz, half a century later, still thought of them as 'a wonderful flight of God's Spirit'.

These troublesome objects first seriously perturbed the world of mathematics in the mid-sixteenth century in connection with the solution of cubic equations. Methods for solving the general cubic equation

$$Ax^3 + Bx^2 + Cx + D = 0$$

and the general quartic equation

$$Ax^4 + Bx^3 + Cx^2 + Dx + E = 0$$

had been developed in the early decades of the sixteenth century by a group of antagonistic Italian algebraists. The methods became widely known through the publication in 1545 of a Latin treatise on algebra by one of their number, Girolamo Cardano (1501–1576), a rather unprincipled genius who taught mathematics and practiced medicine in Milan. He noted that the methods did sometimes run into problems of a nature best illustrated by a simple example. Consider the cubic equation $x^3 = 15x + 4$. This can be solved graphically, as shown in figure 21, by plotting $y = x^3$ and $y = 15x + 4$ and looking for the x-values of the intersection points of these two curves. Evidently, from the figure, there are three such points (or 'solutions'), one positive and two negative. There is, therefore, nothing startling about the problem or 'imaginary' about the solutions. Cardano, naturally, was interested in proceeding algebraically and he argued as follows. Note first the identity

$$(a - b)^3 = -3ab(a - b) + (a^3 - b^3)$$

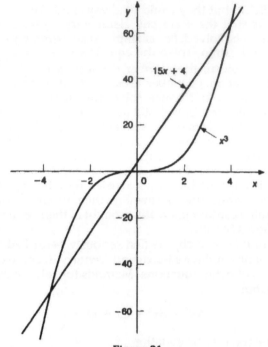

Figure 21

which is readily established by expanding $(a - b)^3$. If we compare
this identity with our equation $x^3 = 15x + 4$, then clearly we can
write $x = a - b$ if only $-3ab = 15$ (i.e. $ab = -5$) and $a^3 - b^3 = 4$.
Substituting $b = -5/a$ into the equation $a^3 - b^3 = 4$ now produces
a quadratic equation in a^3 of the form

$$(a^3)^2 - 4a^3 + 125 = 0.$$

Using the well known formula for solving quadratics now provides the two possible answers

$$a^3 = 2 + \sqrt{-121}$$
$$a^3 = 2 - \sqrt{-121}.$$

In an analogous fashion, by substituting $a = -5/b$ into the
equation $a^3 - b^3 = 4$, we can obtain a quadratic for b^3 with the two
solutions

$$b^3 = -2 + \sqrt{-121}$$
$$b^3 = -2 - \sqrt{-121}.$$

Using these solutions we note that the plus sign of the a^3 expression must be paired with the plus sign of the b^3 expression (and correspondingly the minus sign with the minus sign) in order to ensure that $a^3 b^3 = -125$ (i.e. $ab = -5$) as required. It follows that the solution of our original equation must be

$$x = a - b = (2 + \sqrt{-121})^{1/3} + (2 - \sqrt{-121})^{1/3}.$$

In short, there only seems to be one solution and, to make matters worse, it contains imaginaries. And yet how can this possibly be when we have already seen that there are actually three real number solutions? We can even find their numerical values with a little effort on the calculator. One is exactly $x = 4$, while the others are close to $x = -0.268$ and $x = -3.732$, respectively. What on earth has happened?

Well, those of you familiar with complex numbers (as, of course, Cardano was not) will see through at least part of the 'smoke screen' as follows. We should nowadays write the above solution as

$$x = (2 + 11i)^{1/3} + (2 - 11i)^{1/3}$$

where $i = \sqrt{-1}$. Using regular algebra, together with the condition $i^2 = -1$ where necessary, we readily verify that

$$(2 + i)^3 = 2 + 11i$$

and

$$(2 - i)^3 = 2 - 11i.$$

It follows that our algebraic solution for x can immediately be simplified to

$$x = (2 + i) + (2 - i) = 4.$$

In this fashion we have at least reproduced one of the known solutions. The other two can be generated by realizing that $2 + i$ and $2 - i$ are not the only complex numbers which, when cubed, make $2 + 11i$; but this takes us too far away from our story and there is no need to pursue more details here.

The important point, as far as Cardano was concerned, was the fact that these 'objects' involving the square roots of negative numbers arose in the 'working' necessary to solve equations that had perfectly good real number solutions. It was therefore absolutely necessary to know how to deal with them in a self-consistent fashion, even if their meaning in the context of 'number' remained obscure. The challenge was evident, but it took many more generations of effort to finalize the rules of mathematics necessary for carrying out the mission. Eventually, however, the idea of representing $\sqrt{-1}$ by a symbol, and the algebraic rules for manipulating it, were achieved (primary contributions being made by Leibniz, and later Euler). Euler published an elementary text book in 1768 which discussed in detail how to handle these new numbers, but he had great difficulty in explaining exactly what they were. In it he states 'It is clear that the square roots of negative numbers cannot be reckoned among the possible numbers ... they exist only in our fancy or imagination'. The final problem of conceptualizing these new numbers and of successfully relating them to their forebears was first achieved by a self-taught Norwegian surveyor, Caspar Wessel (1745–1818) in 1797, and published in the *Transactions of the Danish Academy* in 1798.

The answer was in a two-dimensional geometric representation. In this picture a number like $2 + 11i$, for example, was to be pictured as a point in a plane (2 'units' to the 'East' and 11 units to the 'North') rather than as a point on a line, as a real number was. The real number system is then contained within its 'complex' number extension as points along the East–West axis. Unfortunately for Wessel, the *Transactions of the Danish Academy* were not a common source of information for the mathematical research world of the day, and his achievement remained unnoticed until a French translation appeared in 1897. Meanwhile, a Swiss book keeper, Jean Robert Argand (1768–1822), who was also an amateur, independently developed a similar representation in 1806, and has since received most of the credit for the innovation.

Although the unfortunate terminology referring to the two axes of the 'complex plane' as real and imaginary persists to this day, there is nothing whatsoever imaginary about $i = \sqrt{-1}$. The perception of number has merely been raised from a one-dimensional quantity to a two-dimensional one. Within the two-dimensional picture, the meanings of addition, subtraction, multiplication and division can be given a simple geometric interpretation. In addition, each operation always produces another complex number so that the entire system is 'self-contained' and, in this sense, properly defines a number field. Even raising complex numbers to complex powers represents no problem within the system. No mysticism remains—only the unfortunate nomenclature which still insists on referring to the complex number $a + ib$ as the sum of a 'real' part a and an 'imaginary' part ib. I suppose the difficulty surrounds the fact that two-dimensional 'complex' numbers require two axes for their representation. One corresponds to the number line of the old 'real' number system (in itself an unfortunate terminology). Realistically, then, what else is there to call the other?

Gauss was aware of this complex-plane visualization of the new numbers and made use of it (in a somewhat disguised form) in his proof of the fundamental theorem of algebra in 1799. This theorem states that an algebraic equation of order n, i.e.

$$Ax^n + Bx^{n-1} + Cx^{n-2} + \ldots + Dx^2 + Ex + F = 0$$

must have exactly n solutions in the new number system. By 1815, Gauss was in full possession of the complete geometrical theory, and he finally published it for all to see in a classical review in 1831. In this review he confronted all logical objection, coined the expression 'complex number', removed all the remaining aura of mysticism, and placed these numbers on as firm a footing, and with as good an objective meaning, as negative numbers.

From this point on, complex numbers began a rapid and triumphant march through every field of mathematics. Of particular note was their invasion of algebra, where the concept of the complex variable $z = x + iy$ soon spawned an examination of the properties of functions $f(z)$ in the complex plane. In fact, the development of the theory of analytic functions of a complex variable (in large measure by Cauchy) is considered by many to be the great masterwork of nineteenth century algebra. Its full impact on the field of physics is perhaps only now being realized. For example, unbelievable as it may seem, the opening sentence of a 1970's text on particle physics states 'The great discovery of theoretical physics in the last decade has been the complex plane'.

Although it is not our purpose to masquerade as a text book it is perhaps pertinent to note the geometric interpretations of addition and multiplication of complex numbers. Working in a purely algebraic fashion, and contracting i^2 to -1 whenever convenient, we find, for example, that

$$(1 + i) + (2 + i) = 3 + 2i$$

and

$$(1 + i)(2 + i) = 2 + 3i + i^2 = 1 + 3i.$$

Geometrically, these operations are shown in figures 22(a),(b). The addition is seen to correspond to a 'completing of the parallelogram'. The multiplication is best envisaged by recasting the complex numbers in the form

$$1 + i = \sqrt{2}(\cos \theta + i \sin \theta)$$

$$2 + i = \sqrt{5}(\cos \phi + i \sin \phi)$$

in which $\cos \theta = \sin \theta = 1/\sqrt{2}$ and $\cos \phi = 2 \sin \phi = 2/\sqrt{5}$. Algebraic multiplication coupled with a little trigonometry then produces the product representation

$$(1 + i)(2 + i) = \sqrt{10}[\cos(\theta + \phi) + i \sin(\theta + \phi)].$$

Thus, if we think of a complex number as a planar quantity possessing both magnitude and direction, then clearly a multiplication of two such numbers implies a multiplication of their magnitudes and an addition of their directional angles. In particular, multiplication by i implies a counterclockwise rotation by 90 degrees. It is then very clear why a multiplication of i by another i produces −1, since the implied multiplication simply rotates the number point from a unit distance 'North' to a unit distance 'West', a point which represents −1 on the real axis.

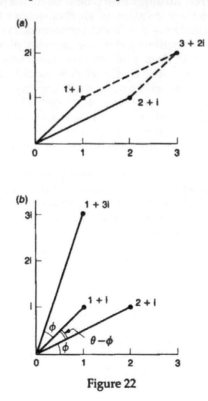

Figure 22

In physics, complex numbers found an early application in equations concerned with oscillatory motion, such as those which occur in association with pendulums. Their usefulness in this context rests upon the beautiful mathematical identity

$$e^{i\theta} = \cos\theta + i\sin\theta$$

relating exponential and trigonometric functions. Based on earlier work by the French born (but English educated) Abraham De Moivre (1667–1754), this celebrated equation is generally attributed to Euler. To illustrate its use, we note first that the differential equation for a completely undamped oscillator is

$$\partial^2 u/\partial t^2 = -\omega_0^2 u$$

in which t is time and u is displacement from equilibrium (or from rest, if you prefer). This equation is readily seen (appendix 2) to be satisfied by any of the solutions $u = \cos(\omega t)$, $u = \sin(\omega t)$ or $u = e^{i\omega t}$, if only the frequency variable ω is set equal to ω_0. In this case, therefore, the complex exponential form is no more convenient than the trigonometric ones.

In real life systems, oscillators tend to be damped; that is, they eventually come to a stop if left long enough to their own devices. The differential equation for motion of this kind contains an extra term in the form

$$\partial^2 u/\partial t^2 + 2f(\partial u/\partial t) + \omega_0^2 u = 0$$

where f is a positive real number with the dimensions of reciprocal time. In terms of the 'trial' sine and cosine forms this equation defies solution. But if we try the complex exponential again, we see that it still solves the equation providing that ω is related to ω_0 and f via

$$\omega^2 - 2if\omega - \omega_0^2 = 0.$$

Solving this quadratic equation leads us to an explicit representation for ω of the form

$$\omega = if \pm \sqrt{(\omega_0^2 - f^2)}$$

assuming, as is often the case, that $\omega_0 > f$. The 'frequency' ω of this damped oscillator is therefore a complex quantity. However, its physical meaning is easy to establish by substituting it back into the initial 'test' function $u = e^{i\omega t}$ when we find

$$u = e^{-ft} \cos [(\omega_0^2 - f^2)^{1/2}t]$$

or

$$u = e^{-ft} \sin [(\omega_0^2 - f^2)^{1/2}t].$$

Since e^{-ft} is a real number which gets forever smaller as time t increases, we see that these solutions describe oscillations that stay constant in frequency but which die away exponentially in amplitude as time progresses.

One of the most important manifestations of wave-like motion in physics is that of the electric field (and accompanying magnetic field) in a light wave. Perhaps the earliest adaptation of complex numbers to problems involving light was that of the Frenchman Augustin Jean Fresnel (1788–1827) in his efforts to understand optical refraction and reflection. Fresnel's theoretical research, carried out in the early 1820s, was based upon a not-wholly consistent elastic–solid theory of light propagation. The complete and self-consistent set of differential equations necessary for describing electromagnetism in general (and the electromagnetic theory of light in particular) were eventually derived by the British physicist James Clerk Maxwell (1831–1879) and published in 1873. These equations, when applied to light, are all of a damped oscillator form and are consequently ideally suited for solution in terms of complex numbers.

Equations of the damped oscillator kind arise time and time again in modern physics since atomic vibrations and molecular vibrations are all of this form (at least in a classical picture). In the equivalent quantum picture, the quantity f (or more precisely its reciprocal) is a measure of the 'lifetime' of the corresponding quantum particle or 'phonon'. Formalisms of this kind became even more important with the coming of wave mechanics and the recognition that all particles of matter, in or out of a solid, possess wave-like features which can be described by a differential equation related to oscillators. This equation, the Schrödinger equation, which permeates all of modern physics at the atomic level (and is set out in more detail in chapter 12) involves complex numbers in such an essential way that it is difficult to see how the subject could have developed at all without them.

Every algebraic equation with real or complex coefficients can be solved in terms of complex numbers. They are therefore really the only basic numerical 'tools' necessary for use in scientific measure. On the other hand, scientists often find themselves using experimental probes and measuring material properties that require far more than a single number (real or complex) to define. The root of the problem is the fact that most experimentation goes on in a world of three dimensions, the laboratory. Common experimental stimuli S and their responses R then tend to be quantities which possess both magnitude and orientation in the three real dimensions of space. Such quantities are called

Josiah Willard Gibbs, 1839–1903. Reproduced by permission of AIP Emilio Segrè Visual Archives.

(three-dimensional) 'vectors'. The properties **P** of matter which these probes and responses measure via an equation of symbolic form $S = P \cdot R$ (examples of which include stresses and strains, polarizabilities and susceptibilities) are then mathematical quantities which are even more complicated than vectors. They are examples of 'tensors'. The algebraic language of tensors, and of their vector generators, has now become a standard language of modern physics. It was pioneered in the 1880s primarily by the American physicist and Yale Professor Josiah Willard Gibbs (1839–1903). A vector is a quantity that has both magnitude and direction. It is numerically determined by its (real or complex) component values along the orthogonal axes of the space in which it is defined. A simple example in three dimensions might be a force F, numerically defined by an ordered trio of numbers

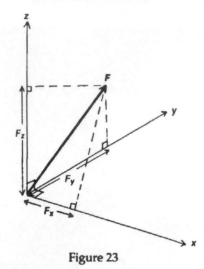

Figure 23

(F_x, F_y, F_z) in the Cartesian coordinate system x, y, z, see figure 23. In two dimensions, a vector is defined by an ordered pair of numbers and obviously has many structural characteristics in common with a complex number. It is important, however, to appreciate that the two concepts are not identical. Just as a point P on a real line can be measured by *any* real number depending on where one arbitrarily defines the origin (or zero) of measure, a two-dimensional vector can be measured by *any* ordered pair of real numbers (this time depending on the directions chosen for the axes as well as the position chosen for the origin). Thus the vector exists independently of any particular set of numbers used to label it.

As an introduction to the language of vectors and tensors let us consider Newton's law, force equals mass times acceleration; say $F = \mathbf{M} \cdot A$ in symbols, where F is the force vector set out above, $\mathbf{M}$ is mass, and A is acceleration. Acceleration, which is the rate of change of velocity, is also a vector quantity (possessing magnitude and direction). In three dimensions it can therefore be expressed in the manner (A_x, A_y, A_z). It is usual to think of mass as capable of being measured by a single real number (say n grams) but, as we shall see, in its most general guise (and particularly, for example, for electrons in semiconductors) it can be something more formidable than that. Newton's law states that each of the components of F is linearly related to the components of A. The full component relationship between force and acceleration in the x, y, z coordinate system then takes the form

$$F_x = M_{xx}A_x + M_{xy}A_y + M_{xz}A_z$$
$$F_y = M_{yx}A_x + M_{yy}A_y + M_{yz}A_z$$
$$F_z = M_{zx}A_x + M_{zy}A_y + M_{zz}A_z.$$

In 'shorthand' we can write this, for convenience, as

$$\begin{pmatrix} F_x \\ F_y \\ F_z \end{pmatrix} = \begin{pmatrix} M_{xx} & M_{xy} & M_{xz} \\ M_{yx} & M_{yy} & M_{yz} \\ M_{zx} & M_{zy} & M_{zz} \end{pmatrix} \begin{pmatrix} A_x \\ A_y \\ A_z \end{pmatrix}$$

so that it looks like a multiplication ($F = \mathbf{M} \cdot A$) with the above pattern of numbers defining the meaning of the 'dot' product. Clearly, if F and A are vectors, then $\mathbf{M}$ is something more complicated since it requires not 3, but $3^2 = 9$, components to describe it. It is called a second-rank tensor—second rank because it has 3^2 components. In this same terminology a vector, with 3^1 components, could also be called a first-rank tensor, and any property which can be completely defined by $3^0 = 1$ component could also be thought of as a tensor of the zeroth rank, although it is more commonly referred to as a 'scalar'.

It follows that mass, in general, is a second-rank tensor, and requires nine components or tensor 'elements' to completely describe it. For electrons moving under the influence of force in a semiconductor much of this complexity is actually present, although not all the elements are fully independent. For the much more familiar case of, say, a pitcher throwing a baseball, all the 'off-diagonal' elements M_{xy}, M_{xz}, M_{yx}, M_{yz}, M_{zx} and M_{zy} are zero, and all the 'diagonal' elements M_{xx}, M_{yy} and M_{zz} are equal (say, to M). It follows that for this case only a single number M (in some appropriate units) is required to define mass, and this accounts for our everyday inability to appreciate the fact that mass is really a second-rank tensor.

Rules for determining how the components of a vector and the elements of a tensor change under rotation of the axes x, y, z are not difficult to formulate and a new mathematics of vector and tensor analysis begins to emerge. Within it, tensors are readily defined to arbitrarily high ranks, and those up to fourth rank are commonly studied in solid state physics. The latter have $3^4 = 81$ components or elements and can be quite formidable 'objects' to measure in detail, although symmetry (either inherent in the physics of the property being measured or in the structure of the

material under study) often restricts the number of truly independent quantities necessary for their definition. Nowadays, particularly in the study of crystalline properties, the Cartesian
subscripts x, y, z are usually replaced by a numerical nomenclature 1, 2, 3, the axis directions to be associated with these
numbers following an accepted convention related to the principal axes of symmetry of the crystal structure in question.

The most common scalar quantities in physics are frequency,
volume, time and temperature. Common physical quantities that
are vectors (in addition to force and acceleration as mentioned
above) include magnetization and polarization, magnetic and
electric fields, electric current, atomic displacements and velocities. Second-rank tensors arise most often as the 'ratio' of two
vectors. Typical examples are stress (defined as the ratio of force
to area), strain (the ratio of two atomic displacements), electric
and magnetic susceptibilities (the response of polarization and
magnetization to electric and magnetic fields, respectively) and
electrical conductivity (electrical current response to an applied
electric field). Each has nine tensor elements, each of which can
be a real or complex number when expressed in appropriate units
of measure.

Third- and higher-rank tensors also make frequent appear-
ances in solid state physics. A common third-rank tensor, with
$3^3 = 27$ elements, can be defined as the response of electric
polarization P (or charge displacement per unit volume) to stress
$\mathbf{X}$. Polarization, a vector, can be specified by its three components
P_i ($i = 1$, 2, 3) and stress, a second-rank tensor, by its nine
components X_{jk} ($i, j = 1, 2, 3$), so that the defining equation for the
response in question becomes

$$P_i = d_{ijk}X_{jk}$$

where repeated indices are summed over all their allowed values.
This third-rank tensor d_{ijk}, called the piezoelectric tensor, is an
important property of some crystal structures. It enables electrical response to be induced by stress. We say of 'some' crystals
because d_{ijk} is identically zero (i.e. all 27 elements are equal to
zero) for many crystal structures by virtue of their high symmetry. The most frequently encountered example of a fourth-
rank tensor, with 81 elements, measures the stress response X_{ij} of
a crystal to an applied strain x_{kl} and therefore appears in a
defining equation of the form

$$X_{ij} = c_{ijkl}x_{kl}$$

where c_{ijkl} is called the elastic stiffness modulus. In spite of its complexity, this tensor plays an extremely important role in the theory of elasticity in solids. Nevertheless, the process of measuring all the independent elements of stiffness for low-symmetry crystals can be quite arduous. For crystals of higher symmetry the task is somewhat easier since the number of independent elements may then be much fewer (as few as three, in fact, for the highest-symmetry cubic structures).

Since, as we have seen above, the ratio (or division) of two vectors is not a vector, it is clear that vectors in three dimensions do not form the basis for a number field. However, the algebra which governs them, and which was pioneered by Gibbs in the 1880s, was actually derived from earlier efforts to extend the concept of a number field to dimensions beyond the two needed for complex numbers. In going from one-dimensional real numbers to two-dimensional complex numbers, all the familiar rules of common everyday algebra are retained. For example, rules like

$$a + b = b + a$$
$$a(b + c) = ab + ac$$
$$ab = ba$$

are equally true whether a, b and c are real or complex quantities. The question of whether the concept of number can be extended to three dimensions was first taken up in earnest by the British School of Mathematics in the early decades of the nineteenth century. Foremost among the early pioneers were the Cambridge algebraist George Peacock (1791–1858) and London University Professor Augustus de Morgan (1806–1871)—the latter of whom could not hold an academic position at either Oxford or Cambridge because of his refusal to submit to the then-required religious test! Their joint conclusion was, rather disappointingly, that algebras in three (or more) dimensions could not be consistently developed. And in a pedantic sense they were correct. The failure was due to an insistence on trying to maintain *all* the rules of complex algebra as if these were in some sense sacred. The idea that such rules and definitions were in fact rather arbitrary, and that other self-consistent algebras could be constructed if only some of these rules were relaxed, occurred first to William Rowan Hamilton (1805–1865), a Fellow of Trinity College, Dublin, and Ireland's most famous mathematician.

Hamilton considered first the obvious extension from two-dimensional numbers $a + ib$, with $i^2 = -1$, to three-dimensional

Sir William Rowan Hamilton, 1805–1865. Picture by C Grey, etched by J Kirkwood. Reproduced by permission of Mary Evans Picture Library.

'numbers' $a + ib + jc$, with $i^2 = j^2 = -1$. Geometrically this looked fine, with i^2 representing a rotation of 180° from $+1$ on the real axis (say the x-axis) to -1 on this same axis by way of the y-axis, and j^2 representing a similar 180° rotation, but this time passing through the z-axis. In this picture there is no difficulty in defining addition by writing

$$(a + ib + jc) + (a' + ib' + jc') = a + a' + i(b + b') + j(c + c')$$

which is just a generalization of the parallelogram law for complex numbers. The big problem arises in trying to devise a geometrical scheme to represent multiplication in terms of three-dimensional rotations. The difficulty centers on the fact that, whereas it takes only two real numbers to define a multiplication in two dimensions (namely the angle of rotation and the ratio by which the length of the vector is changed), it takes more than three real numbers to define an equivalent operation in three dimensions. Specifically, one must define the direction of the axis in space about which the rotation is to take place (which takes two numbers, a latitude and a longitude) together with the angle of rotation (a third number) and the magnitude 'stretch ratio' (a fourth). In particular, writing

$$(a + ib + jc)^2 = a^2 - b^2 - c^2 + 2iab + 2jac + 2ijbc$$

can never be 'correct', regardless of how you try to define the meaning of ij, because it contains only three independent variables a, b and c, and four are needed.

Finally, one day in 1843, as he was walking with his wife along the Royal Canal in Dublin, the inspiration came to Hamilton that he must use number quadruples $a + ib + jc + kd$ with $i^2 = j^2 = k^2 = -1$, but abandon the commutative law for multiplication (which in ordinary algebra, for example, says that 5 times 3 must be equal to 3 times 5). Hamilton realized that he should let ij = k but require ji = $-$k; and similarly for the other 'imaginaries' in the form jk = i, kj = $-$i, ki = j, ik = $-$j. He stopped in his tracks and, with a knife, he cut the fundamental formula

$$i^2 = j^2 = k^2 = ijk = -1$$

on a stone of one of the bridges across the canal.

Just as Lobachevsky was able to create a new geometry consistent within itself by abandoning the parallel postulate of Euclidean geometry, so Hamilton created a new algebra, also consistent within itself, by discarding the commutative postulate for multiplication. Except for the loss of this commutative law, these new numbers, or 'quaternions' as they are called, obey all the normal laws of algebra. The loss of the commutative law for multiplication is not, as you may think, a shortcoming of the resultant algebra. Rather it is the manifestation of a property of rotations in three dimensions that is absent from their two-dimensional analogs. The fact that ordinary complex numbers obey the commutative law in question is just a mathematical statement of the fact that rotations in two dimensions are commutative. That is to say, if you perform two consecutive rotations (R_1 and R_2), it does not matter at all in which order you perform them. A little experimentation will quickly convince you of this fact. On the other hand, in three dimensions things are very different. For example, if you hold this book out horizontally in front of you and let R_1 stand for a rotation of 90° away from you and R_2 stand for a rotation of 90° anticlockwise, you may easily verify a breakdown of commutation for rotations by first performing these rotations in the order R_1R_2 and then again in the reverse order R_2R_1. This breakdown of commutation can be demonstrated for all dimensions higher than two, and algebraic laws must reflect this fact.

Hamilton was an inveterate scribbler. He usually carried pencils and paper wherever he went and, it is said, was occasionally seen to jot notes down on his fingernails and even (according to his son) on his hard-boiled egg at breakfast. Hamilton was convinced that in quaternions he had found the natural algebra of three dimensions. Its four-dimensional form seemed to him to be much more fundamental than any three-dimensional vector coordinate system because operations with quaternions were completely independent of any coordinate representation. On the other hand, the existence of four axes caused difficulties of visualization (at least for scientists) and quaternions did not turn out to be the enormous success that Hamilton hoped they would be. To ease their interpretation Hamilton tried to separate the quaternion representation $a + ib + jc + kd$ into what he called a scalar part (a) and a vector part ($ib + jc + kd$), but the multiplication of two vector parts according to the rules for quaternions always 'coupled-in' the scalar part again. In retrospect, it is clear that the important advance does not center upon quaternions themselves, but on the idea that mathematics is free to build new algebras which need not satisfy the rules of real number or complex number algebra.

Hamilton was not the only person working on vector systems in the mid-nineteenth century. The German mathematician Hermann Grassmann (1809–1877), working completely independently of Hamilton, in 1844 introduced the concept of an n-dimensional vector space and stressed the development of an abstract science of 'spaces' which would include the geometry of three (and two) dimensions as special cases. Grassman laid stress on the different kinds of multiplication that could be defined in spaces of arbitrary dimension and on their geometrical interpretation, particularly for special cases where they reduced to familiar operations. The derivation of vector algebras in higher dimensions was also championed by the English mathematician William Kingdon Clifford (1845–1879) whose 'Clifford algebra' is today being touted by some as the algebra best suited for doing physics. Historically, however, it was the algebra limited to vectors in three dimensions which, under the sponsorship of Josiah Willard Gibbs, won the popularity contest in so far as scientists were concerned.

The vector analysis of Gibbs appeared in 1881, and by the turn of the century vectors (and their generalization, tensors) were beginning to infiltrate the physics world and even to appear in text books. Today, vector algebra (and particularly its differential extension) permeates all of physics. Its advantage, of course, is visualization, since three-dimensional vector quantities (representing forces, velocities, and the like) are explicitly contained in the format. It also enables the equations of physics (such as Maxwell's electromagnetic equations) to be set out in a form that does not require reference to any specific coordinate system (Cartesian, polar, cylindrical, or whatever). The price to be paid for this convenience is the learning of a few special rules that define how to combine vectors in a self-consistent fashion. It is only necessary to compare a 'pre-vectorian' text book on, say, electromagnetism, with its modern counterpart to appreciate the enormous simplification of notation resulting from the use of vector and tensor algebra. And what about quaternions? Although they are rarely mentioned by name today in a scientific context they do have a respected place in algebra. Moreover, although Hamilton's name is not usually associated with vectors, it should not be forgotten that the primary properties of vectors had been worked out by Hamilton in his investigation of new algebras long before Gibbs took up their cause.

Finally, perhaps, a few words should be said about the limitations of vector and tensor algebra as it is presently constituted. A major problem is that not all vectors are alike in their transformation properties in space if reflection is included. Thus, while a velocity v or electric field E change sign upon reflection in a plane normal to their direction, a magnetic field H does not. A magnetic field has the symmetry of a loop of current-carrying wire within the plane normal to H and so stays invariant under the stated reflection. It is therefore necessary to distinguish between 'proper-vectors' (v and E) and so-called 'pseudo-vectors' H. In the same way, since spatial objects can possess a handedness, not all scalars have the same symmetry properties either; some (proper-scalars) remaining unchanged under inversion $(x, y, z) \rightarrow (-x, -y, -z)$, others (pseudo-scalars) changing sign. In vector and tensor algebra these pseudo-quantities have to be given special treatment and be carefully distinguished from their proper counterparts. A vector algebra capable of dealing with all these subtleties would have to be set out in an eight-dimensional space having one scalar, one pseudoscalar, three vector and three pseudovector dimensions. In fact, just such an algebra already exists, called Clifford algebra, after the William Clifford mentioned earlier. It has its proponents, and a number of recent text books (including ones on classical mechanics and electrodynamics) have appeared using the new algebra, but it seems doubtful whether another algebraic revolution is likely to occur in this context at any time soon.

One other immensely important algebra also had its origins in quaternion theory; that is, 'matrix algebra', originated by the Cambridge mathematician Arthur Cayley (1821–1895) when he started to use square arrays of numbers (like that set out for the mass tensor above) to describe rotations. The work grew out of a memoir in 1858 on the theory of geometric transformations. Using these number arrays (which he called matrices) he was able to define operations using them that correctly described the combined effects of two or more consecutive rotations (including their non-commutation in three dimensions) and which determined how vector and tensor elements changed upon rotation of axes. With appropriate definitions it is then possible to think of operations on matrices as constituting an 'algebra', and the study of this matrix algebra has been a leading factor in the development of an increasingly abstract view of algebra in the twentieth century.

Most importantly, matrix algebra proved to be just the mathematical tool needed for the development of many aspects of quantum mechanics. Most famous among these was the invention of 'matrix mechanics' by the German physicist Werner Heisenberg (1901–1975) in 1925. Since the quantum atom, for example, can only have discrete energy states, it is possible to think of these states as defining a 'vector' in an abstract 'space'. The various physical quantities of interest then find their representation as square arrays of numbers (i.e. matrices) in this space, and operations concerning them then become manifestations of matrix algebra. In today's text books on quantum physics the methods of matrix algebra have spread into virtually every corner of the subject. How many of today's quantum physicists recognize that they are standing on the shoulders of Cayley and Hamilton I can only guess but, recognize it or not, all their matrix techniques have grown out of a persistent search for the meaning of 'number'.

6

From Tiling Floors to Quasicrystals

No-one ever suggested that tiling floors with equal-sized square tiles was either difficult or intellectually stimulating. This particular solution to the problem of 'tiling the plane' must surely have been self-evident to the first human who ever gave it a thought. At the other extreme, the general problem concerning which shapes of identical tile can or cannot be fitted together to fill an infinite plane without gaps is one that is still far from a complete solution. Although many solutions other than the one involving square tiles are known (some even to creatures of supposedly lesser intellectual capacity than ourselves, such as bees and wasps in their honeycomb apartmental constructions), tiles can come in so many different weird and wonderful shapes that it is difficult to know how to begin an assault on the completely general problem.

Let us start by restricting our thoughts only to tiles with straight sides. This still involves choices of enormous complexity, so that we first restrict ourselves even further to focus upon tiles of *regular* polygonal shape; that is, tiles with n sides of equal length and n corresponding interior angles of equal magnitude. Listing these in order we start with the equilateral triangle ($n = 3$), square ($n = 4$), regular pentagon ($n = 5$), regular hexagon ($n = 6$), and so on with n advancing arithmetically onward and upward without limit. We may refer to these tiles as the regular n-gons, with the ∞-gon (or infinity-gon) being a circle, a shape which very obviously does not successfully tile the plane. By looking mathematically at the manner in which regular polygons must fit together exactly about a corner point of contact in order to be candidates for tiling, it is a very simple exercise (as shown in chapter 2) to eliminate all but the triangle, square, and hexagon. It

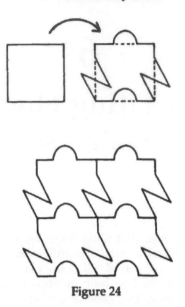

Figure 24

is then equally easy to verify by trial that each of these three remaining candidates can indeed tile the plane in a very simple fashion, as shown in figure 2 of chapter 2. However, even in glancing at these extremely simple patterns it is already possible to sense the enormous complexity of the general tiling problem since we can bend or 'kink' each of the sides of the square tiles, for example, in any fashion whatsoever (so long as parallel sides are treated equally) and obtain arbitrarily grotesque identical tiles that still fit together quite happily to tile the plane. One such example, looking like an army of space invaders, is shown in figure 24. Although some basic rules (called Conway criteria, after their proposer John H Conway) have been formulated to help decide which subsets of arbitrarily grotesquely shaped tiles just might pave the plane, they are less than rigorous and are limited to periodic arrangements. A recent investigation of the general problem concludes that there seems to be no hope for a test which decides whether a given tile can pave in some haphazard manner.

Given this rather sobering conclusion, it is necessary to proceed with our investigation by returning to our restriction to tiles with straight edges (polygons) but now attempt to eliminate the requirement that these polygons be regular. Looking for a simple place to start, we focus first upon the simplest of all polygons, the triangle, and ask the question 'Which triangles can exactly tile the infinite plane without gaps?' If I tell you the answer—that, in fact, they all can (no matter what their shape)—and that the proof is almost trivial, do you think that you can demonstrate it? What we are claiming is that *any* set of identical triangular tiles (long skinny ones, short fat ones, it doesn't matter which) can be 'jigsaw-puzzled' together to exactly fill up the infinite plane without gaps or overlaps. Take a few moments to think about it, or draw a few pictures. Can you find a proof? Well, the flash of insight, or the 'aha' of the proof (as Martin Gardner, the Dean of Recreational Mathematics, would call it), is the realization that any two identical triangular pieces can always be fitted together to form a parallelogram in the fashion shown in figure 25(a). Parallelograms can always be assembled side-by-side to make up endless strips, and finally these strips can then always be stacked together to completely fill up the plane as shown in figure 25(b). In fact, figure 25 is essentially a 'proof without words' of the existence of triangular tiling for the general case.

So much for the triangle! Impressed by the simplicity of the solution, we are now encouraged to soldier on to more ambitious projects. The next simplest polygonal tile is the quadrilateral, with four sides, but this time allowing arbitrary side lengths and angular connections. It is with four-sided figures that we first meet the notion of convex and non-convex polygons. Convex polygons have all their inside (or interior) angles less than 180 degrees or, what is the same thing, π radians. Since the sum of the interior angles of a triangle is π, all triangles are convex by definition. But such is not necessarily the case for quadrilaterals, or indeed for any polygon with a number of sides greater than three. Convex polygons are generally very much simpler to study than their non-convex counterparts. This is particularly so for polygons with a large number of sides since, for them, departures from convexity can occur in an enormous variety of extremely complicated ways, while convex polygons always retain at least some semblance of 'disc-like' simplicity. In the context of quadrilaterals we can perhaps visually associate convex examples with 'kites' and non-convex ones with 'darts' in the manner made clear in figure 26(a).

We now ask 'Which quadrilaterals can tile the plane?' The answer, once more, is that they all can—convex, non-convex and of arbitrary shape. The technique of the proof parallels that for the triangle and can also be expressed in a wholly pictorial fashion as shown in figure 26. For example, take any two quadrilaterals of identical shape and join them along a common edge to

(a)

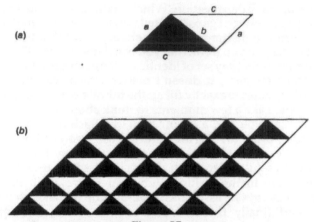

(b)

Figure 25

form a hexagon with three pairs of equal length and parallel opposite sides. Hexagons of this kind will always fit together perfectly by translation alone (that is, without rotation or turning over) and can be readily arranged to tile the plane in the manner made clear by figure 26(*b*) for convex quadrilaterals and by figure 26(*c*) for their non-convex counterparts.

It is now tempting to ask how much longer these almost trivial demonstrations of tiling can last as the number of sides of the polygon is further increased. The answer, unfortunately, is no longer at all. Even if the focus is restricted to convex polygons alone, the problem for the pentagon is a daunting one. First, since the regular pentagon does not tile the plane, as was shown in chapter 2, it is immediately clear that not all pentagons will suffice. However, it is equally clear that some solutions do exist for the five-sided tile, since some successful examples are very easy to create by a closer examination of the honey-bee's solution of the regular hexagonal tiling problem (shown in figure 2 of chapter 2). For example, we can always slice a regular hexagon into two equal halves, each of which will be a pentagon and therefore will immediately give rise to a pentagonal tiling pattern. In fact, the 'bisection' of a regular hexagonal tile can be carried out in an infinite number of different ways so that we have here hit upon an entire class of pentagonal tiling solutions.

An attack on the *general* problem for convex pentagons, unfortunately for us, requires resort to higher mathematics. This was first seriously undertaken by K Reinhardt of the University of Frankfurt, and was reported in his doctoral thesis in 1918. Rein-

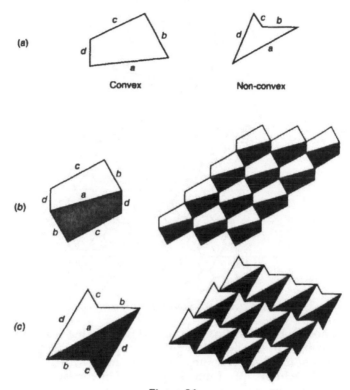

Figure 26

hardt found five different classes of convex pentagon which could perfectly tile the infinite plane. If we label the general pentagon with sides a, b, c, d, e and corresponding angles A, B, C, D, E (as shown in figure 27(a)) then these classes are defined as follows:

(1) $A + B + C = \pi$;

(2) $A + B + D = \pi, a = d$;

(3) $A + B + D = \pi/3, a = b, d = c + e$;

(4) $A + C = \pi/4, a = b, c = d$;

(5) $A = \pi/6, C = \pi/3, a = b, c = d$.

Note that the bisected honey-bee examples are all rather special cases of class 1.

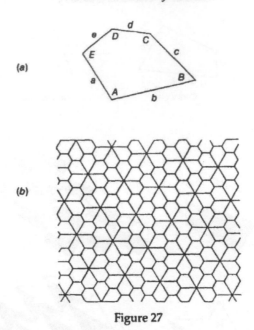

(a)

(b)

Figure 27

Although Reinhardt never claimed that his list was complete, most people in the field thought that it probably was, so that it came as something of a surprise when, nearly forty years later, R B Kershner of Johns-Hopkins University, Baltimore, announced the discovery of not one but three more classes as follows:

(6) $A + B + D = \pi$, $A = 2C$, $a = b = e$, $c = d$;

(7) $2B + C = 2D + A = \pi$, $a = b = c = d$;

(8) $2A + B = 2D + C = \pi$, $a = b = c = d$.

Try drawing a few for yourself and fitting them together. Most of the resulting tilings can be formed into small units of from two to six tiles which then repeat themselves endlessly by translation alone. In figure 27(b), as a rather pretty example, we show a high-symmetry form which is, in fact, a special case of all the classes 1, 5 and 6. Its translational unit, which contains six pentagonal tiles, has a rather appealing flower-like appearance.

The tiling problem for the six-sided convex hexagon, for which the bees have already found the most symmetric answer, was also discussed exhaustively by Reinhardt in his 1918 doctoral dissertation. This case turns out to be rather simpler than that for the convex pentagon, and Reinhardt showed that there are only three different classes of solution as follows:

(1) $A + B + C = \pi, a = d$;

(2) $A + B + D = \pi, a = d, c = e$;

(3) $A = C = E = \pi/3, a = b, c = d, e = f$.

The regular honeycomb lattice is a special case of all three classes.

For convex polygons with more than six sides the problem, somewhat surprisingly, becomes much less difficult since a proof has been known for quite some time that no such polygon exists that can pave the plane in the manner required. The proof isn't trivial by any means, but neither is it of great complexity. Not only will no polygon with more than six sides work, but it has much more recently (1978) been shown that not even an infinite variety of convex polygons, each with more than six sides, can do the job if they are all of finite area and diameter. It follows that our knowledge concerning the paving of the plane is probably complete in so far as identical convex polygons are concerned. Unfortunately, the corresponding problem for non-convex polygons is still far from complete. Even the subcategory of polygons called polyominoes (which are tiles formed by joining together a set of equal-sized squares edge to edge so that the corners match) by themselves give rise to a host of unsolved tiling problems.

Most of these polyominoes are non-convex objects (the complete collection out to pentominoes is shown in figure 28) and, while many can tile the plane, many others cannot. The first difficulty is the fact that there are a bewildering number of polyominoes to consider, even if you restrict your efforts to *n*-ominoes for which *n* is relatively small. For example, while there are 'only' 12 *n*-ominoes with $n = 5$ (the pentominoes of figure 28), there are no less than 4655 polyominoes with $n = 10$ and 3 426 576 with $n = 15$. The associated tiling problems can easily be imagined. Moreover, the general polyomino problem is far worse since there is no restriction at all on the number of sides of a non-convex polygon that can successfully tile the plane. The proof of this fact is almost trivial since it is very simple to divide a square into two identical polyominoes with arbitrarily large *n* (by drawing a stepped 'zigzag' cut down a diagonal of the square with as many steps in it as takes your fancy). Then, since the squares can tile the plane, so also can these polyominoes. At the other extreme we can ask what is the smallest-*n* polyomino that cannot tile the plane? Can all the pentominoes of figure 28 do the trick? Unkindly, I leave it to you to decide. Since it is obvious that the

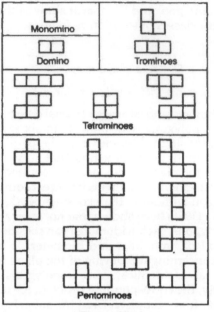

Figure 28

octomino made up of a three-by-three square with the center removed will have no success, the smallest polyomino 'failure' must evidently have *n* less than or equal to eight.

From a more global point of view, the most simple tilings of the plane are those which are periodic in the sense that one can outline a region of tiles that pave the plane by translation alone. But must tilings necessarily be of this form? It is very easy to show that they need not by constructing a simple counter-example. To accomplish this we take a checkers board tiling pattern of squares and bisect each square into two equal-sized rectangles, but we do this by using 'bisecting cuts' which run 'north–south' and 'east–west' in a random manner (see figure 29). While it is true that some shapes, such as squares and regular hexagons, can tile *only* periodically, others are certainly capable of tiling in a non-periodic fashion. However, one quickly finds by trial and error that the shapes that can tile non-periodically seemingly can always be rearranged to tile periodically as well. This is obviously the case for the example shown in figure 29. In fact, it was long thought that a much more general statement was true; namely that even if two or more *different* tiles are used to form a non-

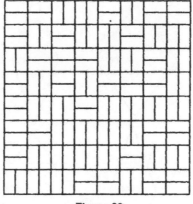

Figure 29

periodic tiling, then they also can be regrouped in some fashion to tile periodically as well (patterns of this form, composed of only a few different tile shapes, being commonly referred to as 'mosaics').

This last conjecture was finally proved to be false in 1964, when Robert Berger, of Harvard University, actually succeeded in constructing counter-examples (albeit very complicated ones with extremely large numbers of inequivalent tiles). Within a few years other researchers succeeded in producing simpler examples until the number of inequivalent tiles required for such a construction was progressively reduced to six, then to four, and finally to two. It is these final 'aperiodic' patterns made up of two inequivalent polygons upon which we shall now focus our attention. First, however, it is pertinent to note that the question of whether there is a *single* tile that can tile in a necessarily aperiodic fashion remains unanswered. It seems unlikely and it is known that it certainly cannot be a convex polygon.

The production of a two-tile necessarily aperiodic tiling of the plane was first achieved by Roger Penrose, a mathematical physicist who holds the Rouse Ball Professorship of Mathematics at Oxford University. The simplest forms that the two tiles can take are rhombuses; that is, parallelograms with four equal side lengths. We show them in figure 30(*a*). One has internal angles of 36 and 144 degrees (the 'thin' tile) and the other has internal angles of 72 and 108 degrees (the 'fat' tile), and they are now universally known as 'Penrose tiles' in honor of their discoverer. In order to *force* aperiodicity it is necessary to restrict the manner

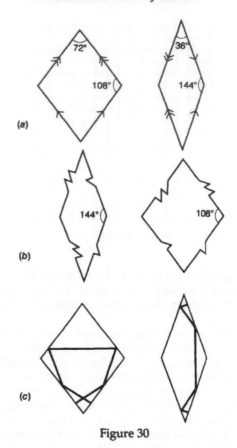

Figure 30

in which these tiles are allowed to be fitted together, the simplest rule being one which matches the like arrows (see figure 30(*a*)) on superposed, or joined, edges. Since this 'matching arrow' rule can readily be imposed physically upon the tiles by deforming the edges with interlocking notches and tabs (as shown in figure 30(*b*)) the true aperiodic tiles should therefore be slight modifications of the 'parent tiles' of figure 30(*a*). However, since appropriate perturbations of the parent tiles can be given in a countless number of forms, discussions are usually presented in terms of the parent tiles together with some arrow-like decoration.

There are an infinite number of different aperiodic tiling patterns which can be composed from the decorated Penrose tiles. One example is shown in figure 31(*a*). The patterns seem to be a

Roger Penrose, 1931–. Rouse Ball Professor of Mathematics, University of Oxford. Courtesy of R Penrose.

(a)

(b)

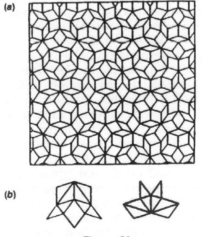

Figure 31

mystifying mixture of order and unexpected deviations from order. As they expand they seem to be striving to set up a periodicity without ever quite managing to do so. Their fascination was such that Penrose first thought of a possible 'application' in terms of artistry or even wallpaper. Little did he know at the time that their fascination would also be one which nature itself would not have the power to resist. Technically, Penrose was able to show that every finite region in any such 'Penrose pattern' is contained infinitely many times elsewhere, not only in that particular pattern, but also in every other Penrose tiling pattern. He was also able to establish the fact that they could be continued to infinity. This he did by deriving a set of 'deflation rules' (shown in figure 31(*b*)) which are rules for dividing each of the Penrose tiles into smaller tiles of similar shape and tile-matching rules. Deflating each tile in a Penrose pattern then creates a new Penrose pattern, but with a larger number of smaller tiles. This process can obviously be continued endlessly and thereby establishes that any such pattern can be continued to 'tile the infinite plane'. The deflation rules also provide a method for calculation of the ratio of the numbers of fat and skinny tiles in the infinite pattern. This ratio is the 'golden ratio'

$$\tau = (1 + \sqrt{5})/2 = 1.618\,033\,989\ldots$$

with there being more fat tiles than skinny ones.

Although there was no realization of the fact at the time, the advent of the Penrose aperiodic lattice was destined to have an immense impact upon crystallography some ten years later. Crystallography, or the study of the positions of atoms in crystals, was touched upon in chapter 2, where we defined a Bravais lattice as being a three-dimensional array of identical atoms (or points), each of which has an identical environment. Such lattices are necessarily periodic and can equally well be defined in two dimensions where, it is found, they can possess two-, three-, four- or six-fold rotational symmetry about a point, but not five-fold symmetry. Examples of two-dimensional 'Bravais' lattices with two-, three-, four- and six-fold rotational symmetries are readily constructed by tiling the plane with rectangles, regular hexagons, squares and equilateral triangles, respectively, and placing atoms at the points where three or more tiles join. An important feature of the Penrose pattern that you can see quite clearly from the example shown in figure 31 is that it has embedded within it a subtle five-fold rotational symmetry. It is not a true five-fold symmetry, since rotating the entire pattern by

360/5 = 72 degrees does not exactly return it to its original form. Nevertheless, there is contained within it an infinite number of pentagonal star structures.

But the orientational symmetry of the pattern goes much deeper than this. Every tile in the pattern, without exception, has sides oriented along one or other of two complementary 'star' directions. The pattern therefore possesses a kind of ten-fold 'orientational' order, although clearly it does not have a spatial (that is, translational) order. Indeed, five-fold or ten-fold orientational symmetry is rigorously incompatible with true spatial (that is, translational) order, or translational invariance to give it its proper title. On the other hand, although there is no rigorous spatial order, there does exist something beyond spatial randomness in the translational characteristics of the pattern. For example, if you examine the pattern of figure 31 by viewing it obliquely along the orientational direction of the tile edges (any one of the ten equivalent directions will do) with one eye closed and the other close to the plane of the figure, you will observe some peculiar patterns. There is a tendency (if you overlook a few renegades) for those tile edges parallel to your direction of observation to form wide and narrow strips, although the pattern of strips does not appear to repeat in any obvious periodic sequence. However, if the individual tiles are 'decorated' with straight line segments in the manner depicted in figure 30(c), then an exact and striking pattern appears with the decoration forming five sets of continuous parallel lines (at relative angles of 72 degrees) which run throughout the Penrose pattern. These lines are called Ammann lines (after their discoverer).

Although the Ammann lines are not parallel to the tile edges, their importance is contained in the fact that the spacing between them, viewed along any one of the five equivalent orientations, takes on two, and only two, values that we shall refer to as wide (w) and narrow (n). The ratio of these two widths is found to be equal to the golden ratio τ, but the recurrence pattern of ws and ns is neither periodic nor random. It has what is known as a 'quasiperiodic' order, and a way of seeing exactly what this means is to devise a procedure that generates such a sequence. One method is to set up a repeating (or 'iterative') scheme involving a substitution rule as follows. Starting with the symbol combination nw, we replace n by w, and w by nw, and then repeat this replacement procedure over and over again *ad infinitum*. In this fashion we generate an increasingly lengthy sequence of ns and ws. Carrying out this recipe we obtain the first few members of this sequence as shown below:

nw

wnw

nwwnw

wnwnwwnw

nwwnwwnwnwwnw.

The result, whether of finite or infinite extent, is also sometimes referred to as a Fibonacci sequence, after the thirteenth century Italian mathematician Leonardo Fibonacci who first generated it (although, of course, in a completely different context). It never contains two or more *n*s together, and the *w*s appear only singly or in pairs. The substitution rule $n \to w$, $w \to nw$, is a sort of one-dimensional inflation rule, the reverse of which ($w \to n$, $nw \to w$) is a corresponding deflation rule which enables us to 'test' any likely looking sequence of *n*s and *w*s for quasiperiodicity. Thus, the spacing of the Ammann lines, and therefore the fundamental translational 'periodicity' of the Penrose pattern, is that of a subtle sequence of wide and narrow spacings which never quite repeats but is decidedly non-random.

This Fibonacci sequence can be generated geometrically in a most intriguing fashion as follows. Take an infinite square lattice of points (n_1, n_2), like the line-crossing points on a boundless checkers board, and on it, in any location whatsoever, draw a straight line of slope $1/\tau$ (figure 32). Did I say *any* location? Well, more precisely any location for which the line does not go exactly

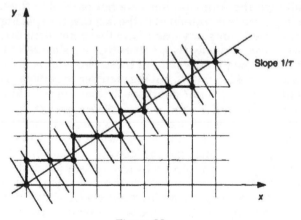

Figure 32

through any of the points. This limitation may at first glance seem
to be almost impossibly restrictive, since the line is to be ideally of
infinite extent. But the restriction is illusory since the slope $1/\tau$ is
an irrational number, and the probability that any such line
chosen at random will pass exactly through a lattice point is, in
fact, infinitesimally small. Now, starting from any point close to
the line (say the bottom-left-hand-corner point in figure 32) draw
a zigzag line of unit horizontal and vertical components according to the single rule that every side is drawn in such a manner as
to keep the zigzag line as close as possible to the straight line. If
horizontal segments are labeled 'w' and vertical segments are
labeled 'n', we are now able to write down the sequence of ws and
ns contained within the zigzag line. Using the specific example of
figure 32 we find the pattern

$$nwwnwwnwnwnwnwnw \ldots$$

and we can easily test it (using our deflation rule) for quasiperiodicity; and quasiperiodic it is! If we now drop perpendiculars
from the zigzag corners of the $1/\tau$ slope line (see again figure 32)
then we generate a set of Ammann lines. It follows, therefore, that
the Ammann lines can be obtained as a 'projection' from a two-
dimensional square lattice onto a particular straight line.

The question now arises as to whether more generalized
projection methods (analogs of that set out above) can be found
in higher dimensions. For example, can the pattern of Penrose
tiles itself be projected out of a set of lattice points (n_1, n_2, ...) of
some space with dimension three or larger? The answer is yes,
and was first demonstrated by the Dutch mathematician N G de
Bruijn in 1981. It turns out that the Penrose pattern can be
generated from a suitably chosen two-dimensional irrational
'slice' of a five-dimensional set of lattice points (n_1, n_2, n_3, n_4, n_5).
But now the question arises as to why we should stop at two-
dimensional projections. What if one selects three-dimensional
irrational slices of even higher-dimensional sets of lattice points?
Can one in this manner perhaps produce a three-dimensional
'tiling' of space which is quasiperiodic?

The possibility of generalizing the Penrose tiling pattern to
three dimensions was a problem considered essentially simultaneously by several mathematicians in the early 1980s, and
success was eventually achieved using the projection method;
specifically by taking an irrational three-dimensional projection
from a six-dimensional set of lattice points. In place of Penrose's
two rhombic building tiles of figure 30(*a*), we now obtain two
rhombohedral building 'blocks' (each with six identical rhombus-

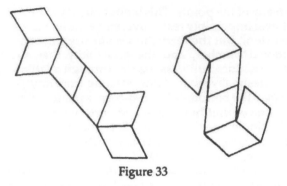

Figure 33

shaped faces), one 'fat' and one 'skinny'. Models of these two rhombohedra can be made by folding up the rhombus 'nets' shown in figure 33. In general, these two kinds of three-dimensional 'parent' tiles can be packed together periodically or non-periodically (just as was the case for their two-dimensional Penrose counterparts) so that additional matching rules are required to ensure orientationally ordered quasiperiodicity—matching rules which, however, are automatically provided by the projection technique. Physically, as was the case in two dimensions, the matching rules can be assured by slightly deforming the parent tiles with notches and tabs. Orientationally, these three-dimensional Penrose analogs possess the symmetry of an icosahedron (figure 1 of chapter 2), a symmetry which for ordered lattices is just as forbidden in three dimensions as were the planar five- and ten-fold symmetries of the Penrose pattern in two dimensions. Once again the golden ratio $\tau = (1 + \sqrt{5})/2$ is found to be intimately involved in all aspects of the new quasiperiodic structure. We shall mention here only one of these aspects, namely that the ratio of the number of skinny tiles to the number of fat tiles in the structure is $1/\tau$.

After the successful mathematical generation of perfect three-dimensional quasiperiodic tilings of space, the question of the possible existence of such patterns for atomic arrangements in actual solids was naturally raised. Could perhaps a whole new branch of crystallography be possible for which orientational icosahedral symmetry was allowed? Penrose himself was skeptical. 'In principle, yes,' he would say, 'but how on earth would nature do it?' What kind of forces in nature could possibly precipitate quasiperiodicity in crystals since, as we must be careful not to forget, without matching rules the 'magic' fat and skinny tiles are also able to tile in a periodic (or even a random) fashion? Even with some physical manifestation of the matching rules being present, perhaps in the form of atoms positioned asymmetrically within the individual fat and skinny tiles, difficulties faced by the growing 'quasicrystal' still abound. The problem confronting nature is not unlike that faced by the 'intelligent layman' who attempts to construct a two-dimensional Penrose tiling starting from a bag of properly tongue-and-grooved tiles such as those of figure 30(b). What happens is that, unless he is very lucky, a first effort to build up the quasistructure will fail, a point being reached (usually after only a modest number of tiles have been set out) where no new tile will 'fit'. Considerable thought and backtracking is then necessary to avoid the geometrical conflict at hand and to 'get it right' before the building process can continue. What sort of forces could possibly be that clever? It long appeared that only long-range forces with almost pathological properties would be capable of such dexterity. However, recently it has been shown that relatively short-range forces of a less pathological kind could possibly achieve the desired result.

Should they exist, the most striking feature of 'crystals' with a three-dimensional quasiperiodic structure (or 'quasicrystals' as they have come to be called) would be the fact that they are not subject to the normal restrictions imposed by the rigorous spatial theorems of Euclidean geometry. Most important among the latter is a restriction, for translationally invariant crystal structures, to symmetry axes with only two-, three-, four- or six-fold rotational invariance. These orientational symmetries are observed experimentally by an indirect, but well understood, technique. The atoms in normal crystals are arranged in families of parallel planes, each of which acts as a mirror to incoming x-rays, electrons or other rays or particles that travel through space as waves. Each plane reflects the incident waves very weakly, but if these weak reflected waves from each member of a family of planes combine in phase, then the total intensity of the reflected wave can become very strong. The phenomenon is called diffraction, and the angles through which a beam of electrons or x-rays are diffracted reveal the symmetry and dimensions of the unit cell of the crystal. However, a quasicrystal does not exhibit equally spaced planes of atoms, so that the diffraction pattern to be expected from such a structure is not immediately obvious and one's first suspicions are that it would not consist of sharp lines at all. Surprisingly, this is not the case. The first theoretical work on the diffraction patterns to be expected from quasicrystals was published in 1982 by Alan Mackay of the University of London. The work was applied to a two-dimensional Penrose pattern (with atoms placed at the corners of each Penrose tile). The conclusion was that the Penrose quasicrystalline diffraction pattern would be just as sharp as for a translationally invariant crystal, but it would have a ten-fold orientational symmetry.

Then, in 1984, just as the theoretical notions of extending Penrose-style quasiperiodic tilings to three dimensions were being developed mathematically, a remarkable observation was reported from the National Bureau of Standards (now the National Institute of Standards and Technology) in Maryland. There, a visiting Israeli scientist, Dan Schechtman, had been pursuing the structure of alloys of manganese (Mn) and aluminum (Al) by electron diffraction techniques, when he stumbled upon a composition, close to Al_4Mn, for which the observed diffraction pattern possessed a supposedly impossible five-fold rotational axis. As so often happens in scientific research, the new material was discovered quite accidentally (as part of a survey to develop lighter and stronger aluminum alloys). This particular composition was prepared by a method known as melt-spinning, in which the melted alloy is squirted onto a cold and rapidly spinning wheel so that it cools and solidifies very quickly, often at rates as fast as a million degrees (centigrade) a second. This abrupt cooling process, called quenching, can often 'shock' materials into unusual structures, and was mentioned in chapter 2 in the context of metallic glass preparation.

In this case, however, the material obtained was clearly not a glass since the diffraction pattern consisted of fairly sharp features. By rotating the sample in the electron beam it was observed that the diffraction pattern had six five-fold symmetry axes, as well as other axes of two- and three-fold rotational symmetry. By measuring the angles between these various axes it was confirmed that the overall rotational symmetry was that of an icosahedron. Within a few weeks of the observation, Dov Levine and Paul Steinhardt of the University of Pennsylvania showed that the observed diffraction pattern was just that to be expected from the simplest form of three-dimensional quasicrystal — namely the so-called icosahedral quasicrystal. It therefore seemed distinctly possible that this particular composition Al_4Mn might be the first known example of a completely new form of matter, existing somewhere between the realms of the completely ordered 'proper' crystal and the completely disordered glassy state. In glass, the atoms possess neither translational nor orientational order, and they usually form (as discussed in chapter 2) when a melt is cooled too rapidly for the ordered crystalline state to assemble itself. When Al_4Mn was rapidly cooled, it should (according to conventional wisdom) have formed either a crystal or a glass depending on the rate of cooling. However, it appeared that Al_4Mn might actually have chosen a new path to solidification somewhere in between a crystal and a glass.

Why the uncertainty, you may be asking, when the observed diffraction pattern was in accord with quasicrystalline expectations? Well, the problem was that the diffraction spots with the peculiar icosahedral symmetry were not quite as sharp as would have been expected for a perfect quasicrystal. Given this situation, the appearance of the icosahedral diffraction pattern might have other possible explanations. First, suppose that the Al_4Mn sample were in reality composed of many ordinary, but extremely tiny, crystals arranged in a specific (and in this particular case icosahedral) orientational pattern. The crystallographers term for this is 'multiple twinning' and, in a more general context, it is not an uncommon occurrence in crystal growth. A second possibility might be that the alloy is an example of a normal ordered crystal with an extremely large primitive unit cell (this unit cell being the smallest crystalline volume that exactly repeats itself to build up the crystal) containing within it one or more orientationally ordered icosahedral clusters of atoms. Third, the alloy might be an example of a phase possessing orientationally ordered icosahedra of atoms, but with a glass-like positional disorder of the individual icosahedra with respect to each other. All of these possible structures would give rise to quasicrystal-like diffraction patterns but with the individual lines (or 'spots') broader than would be anticipated for a true quasicrystal.

As these theoretical ideas were being developed in the late 1980s, materials scientists set about looking for other examples of the quasicrystalline phenomenon by rapidly cooling other alloys of similar chemistry. Success was not long in coming and soon there were dozens of the so-called 'icosahedral' alloys available for experimentation and study. All showed the now-famed 'forbidden' icosahedral symmetry, but all had diffraction patterns with lines that were broader than expected for a perfect quasicrystal, and there was a lively debate in the literature over the relative merits of the various explanations set out above. The primary difficulty seemed to be the fact that these materials, whatever their detailed structure, were probably (like glasses) not in their lowest, or thermodynamically stable, energy state. Prepared, as they were, by rapid cooling, they almost certainly represented states that would (given an eternity of time) transform to their truly stable form which, it was supposed, would be crystalline in the normal sense.

The situation changed suddenly in 1989 when physicists at Tohoku University in Japan discovered a new family of icosahedral alloys that could be prepared by very *slow* cooling of the melt. They were alloys of aluminum (Al), copper (Cu) and either iron (Fe) or ruthenium (Ru) with compositions near $Al_{65}Cu_{20}Fe_{15}$ and $Al_{65}Cu_{20}Ru_{15}$. These quasicrystals were therefore undoubtedly in their thermodynamically stable state and, most significantly, their x-ray diffraction patterns were every bit as sharp as those of regular crystals. They were, therefore, definitely not manifestations of twinning or a glass-like aggregation of icosahedral molecules. Although the possibility of having a normal crystal with an enormous unit cell containing large numbers of icosahedral clusters of atoms can never be *exactly* ruled out (since one can always postulate the existence of a unit cell so large that it is forever just beyond the capacity of experiment to observe) such a situation seems to be far less likely to arise than does quasicrystallinity itself. Furthermore, since almost all the physical properties of the material would still be determined by the quasicrystalline substructure of the ultra-large unit cell, this possible deviation from true quasiperiodicity is really only an academic one. This is particularly true when we recall, from chapter 1, that nature (except at the absolute zero of temperature) will always introduce imperfections into the lattice in some manner. In the context of quasicrystals these imperfections will probably be errors in satisfying the 'matching rules' at some places. These growth 'errors' may actually make it easier for the quasicrystalline system to form at room temperature than at very low temperatures where such imperfections are less tolerated.

In the first preparations of the Al, Cu, and Fe or Ru quasicrystalline alloys the largest 'single crystal' grains obtained were still quite small—often no more than a fraction of a millimetre across. Since that time, however, the experimental situation has improved dramatically and these (and other three-component alloy compositions containing aluminum, copper and lithium, or aluminum, palladium and manganese) can now be prepared in centimetre-sized single-crystal grains by cooling at rates of less than one degree per hour. It seems, therefore, that almost perfect stable quasiperiodic crystals do exist in nature. For centuries scientists believed that every pure solid was necessarily either crystalline or glassy. Now, with mathematicians leading the way, it seems that this view must be changed, and that a conceptually new state of matter needs to be recognized. As 'Murphy's Law' states, 'If it can happen, it will!' The mathematical demonstration of the 'can' this time only just preceded the experimental demonstration of the 'will', but precede it it did. The 'fuzziness' of the diffraction spots in the experiments on the rapidly cooled samples were almost certainly caused by strain defects (sometimes called 'phason' disorder) so that the year 1984, in addition to its Orwellian claim to distinction, will also be remembered as the year of discovery of the quasicrystalline state of matter. But as far as a detailed understanding of this new state of matter is concerned, the work has still barely begun.

As of this writing, a complete knowledge of the precise atomic positions within any quasicrystal remains unknown. While the location of the primary diffraction spectral lines in an x-ray experiment may well be sufficient to determine the quasicrystalline nature of the structure, this only informs us of the geometry and packing arrangement of the two types of tiling blocks. In order to locate the positions of actual atoms within these three-dimensional Penrose cells, it is necessary to interpret both the location and the relative intensities of all the spectral lines. In a real icosahedral quasicrystal each 'fat' or 'skinny' Penrose cell will contain a few or many individual atoms at positions inside it, or on its faces, edges or vertices. It follows that many different quasicrystals can be defined with the very same quasicrystalline arrangement of unit cells. Thus, we do not truly know what the structure is until we know not only what the unit cells are, and how they are arranged in space, but also where the atoms are within the cells. No doubt, before this book reaches publication, at least one such complete determination will have been accomplished. However, one should also bear in mind the possibility that the atoms within the Penrose cells may, in some cases, be so symmetrically disposed that they do not give rise to matching rules. In cases like this, it is possible to construct *random* packings of the cells which still completely fill space without periodicity and still produce sharp icosahedral diffraction patterns.

To this point we have emphasized the *icosahedral* quasicrystal in our discussion of three-dimensional quasiperiodic systems. This has been done primarily because both theoretical and experimental interest first focused on quasicrystals of this kind. However, it is important to realize that many other different quasicrystalline arrangements are mathematically possible in three dimensions, and that some others have already been found in nature. For example, it may have occurred to you that it is possible to envisage a three-dimensional crystal which possesses a true Penrose two-dimensional quasistructure in, say, the xy-(or 'horizontal') planes and a proper periodicity in the z-(or 'vertical') direction. All that is necessary is to have two unit cells with Penrose (figure 30(a)) diamond-shaped tops and bottoms, and 'vertical' rectangular sides. These may then be assembled into a Penrose 'quasilayer', and these layers then stacked on top of each other in a periodic fashion. Such crystal structures have been discovered and are referred to as decagonal quasicrystals.

The first decagonal quasicrystal was identified soon after the discovery of the original aluminum–manganese icosahedral system. It was, in fact, obtained in the very same alloy by cooling at a somewhat slower rate than was necessary to obtain the icosahedral quasicrystal. It was identified by its diffraction spectrum, which had a ten-fold orientational symmetry within 'horizontal' planes but a proper translational periodicity normal to these planes. Naturally, the early samples exhibited the same somewhat broadened diffraction lines that were noted earlier for the icosahedral samples, again due to their metastable thermodynamic nature. Recently, however, stable decagonal quasicrystals have also been obtained by slow cooling (at a rate of a few degrees per hour) in alloys of aluminum, copper, and cobalt (Co) and of aluminum, nickel (Ni), and cobalt. The chemical compositions are close to $Al_{65}Cu_{15}Co_{20}$ and $Al_{70}Ni_{15}Co_{15}$, respectively, and large single grains have now been prepared with sizes up to one cubic centimetre. In samples of this size and quality the diffraction spectra are sharp and the essentially quasicrystalline nature of the structure is confirmed beyond a reasonable doubt. Although the position of the individual atoms within the Penrose unit cells are, at the time of writing, not known with certainty, the atomic 'decoration' shown in figure 34 is thought to be close to the true structure for decagonal $Al_{65}Cu_{15}Co_{20}$. In the figure, the open circles are Al, and the full circles can be occupied by either Cu or Co in a manner determined by the size of the 'sites' (that is, aluminum cages) formed when the cells are assembled into a proper Penrose pattern, a copper atom being bigger than a cobalt one and thereby preferentially populating the larger sites. We note, in particular, that the atomic decoration of figure 34 does exhibit a 'matching rule' and therefore would tend to force a proper Penrose quasicrystalline space-filling rather than a random packing with decagonal symmetry.

As larger and larger specimens of good quality stable quasicrystals become available, experimental studies of quasicrystal chemistry and physics will no doubt advance rapidly. Literally

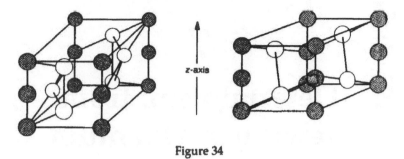

Figure 34

hundreds of compounds have now been observed to have quasicrystalline phases, so that the evidence is increasing that quasicrystals are not at all a rarity. Other classically forbidden orientational symmetries (such as eight-fold and twelve-fold) have also been reported. Most of the materials are metallic alloys, but this restriction may only be from lack of work on other classes of material. At present their potential value for technology is largely speculatory, although they do appear to produce excellent low-friction coatings. Some theorists suggest that, because of the pattern of electronic bonds holding them together, many may become superconductors (that is, materials with zero resistance to electricity) at low temperatures. Others anticipate that their structure will be more rigid than that of ordinary crystals making some of them, perhaps, harder than steel and potentially useful for making superhard tools. Still others believe that many of their properties (their electrical resistance, for example) may exhibit an exotic 'fractal' behavior in which basic patterns are repeated at all scales from the infinitely small to the infinitely large. In truth, no-one yet really knows. Active research continues and, while an understanding of the structural aspects of quasicrystals is now becoming quite advanced, the most important secrets of these intriguing new systems—their unique physical properties—still remain to be uncovered. The only certainty would seem to be that many additional surprises lie ahead; and surprises, after all, are what make science such a captivating endeavor.

7

Determinism: from Newton to Quantum Chaos

Although most of the earliest association between mathematics and physical thinking arose in the context of statics (or, in more common parlance, systems at rest), the description of motion always presented a fascinating challenge and was constantly a subject of scholarly attention. We know, for example, that the early Greeks wondered about the trajectory of a stone when hurled through the air, and also about the orbits of the planets. However, they could see no obvious relation between the two. Such a connection was not made until the seventeenth century, at which time the theory of motion, as caused by forces, was first set out by Isaac Newton in his monumental treatise *Philosophiae Naturalis Principia Mathematica*. This work, first published well over three hundred years ago, manages to sell close to 1000 copies a year to this day.

Anything approaching a full description of the contents of Newton's *Principia* is obviously well beyond the scope of a single chapter in any book, but the spirit emanating from the whole work is that of what we now refer to as 'Newtonian dynamics' and, in particular, of its application (in association with the then newly proposed law of universal gravitation) to the motion of the planets. In *Principia* he formulates the entire program of modern mechanics, first introducing the basic concepts involved (such as mass and momentum) and then setting out his three fundamental 'laws', from which everything else was to follow mathematically once the details of the forces involved were defined.

In the context of planetary motion, the pertinent force is that of gravitation, and Newton stated how this force acts: that every particle of matter attracts every other with a force proportional to the mass of each and inversely proportional to the square of the

Isaac Newton, 1642–1727.

distance between them. When coupled with his second law of motion (namely that the acceleration of a body is proportional to the force acting upon it) this force explains with incredible accuracy a wealth of astronomical observations—from the motion of the planets and comets to the wobbles of the Moon on its axis. Although more than a few philosophical reservations were expressed at the time and in subsequent years (particularly concerning the notion that a force can *instantaneously* act through the vacuum of space over distances of almost unimaginable magnitude), the equations themselves succeeded to a degree beyond anything Newton could possibly have hoped for. In fact, even today, only when approaching the microscopic dimensions of the atom (less than one billionth of a meter) and in the vast reaches of outer space do small deviations from Newton's laws of motion begin to be seen.

Mathematically, Newton's laws of motion are cast in the form of differential equations (specifically, equations which relate quantities to their time dependence). One of the simplest examples can be given in the context of that stone-throwing problem which so interested the early Greeks. If, when a stone is thrown from ground level, we measure the distance 'upwards' as y, and the horizontal distance from the throwing point as x, then Newton's equations make direct statements about the constancy of acceleration in the vertical and horizontal directions as follows:

$$d^2y/dt^2 = -g \qquad d^2x/dt^2 = 0$$

in which t is time and g is the gravitational acceleration constant near the Earth's surface (with a measured value of about 32 ft s^{-2}). Thus, the equations say that the stone experiences a constant *downward* (i.e. negative) acceleration due to gravity for the vertical component of the motion, and zero acceleration in the horizontal direction.

These equations are easy to 'solve' (by which we mean cast in the form y is some function of time t, and x is some other function of t) for anyone having a smattering of knowledge concerning the 'rules of the game' for differential equation analysis. The result is

$$y = vt - \tfrac{1}{2}gt^2 \qquad x = ut$$

where v is the initial upward component of velocity and u is the initial horizontal component of velocity. Plotting points (x, y) as a function of t for, say, $u = v = 16$ ft s^{-1}, we can now sketch the trajectory of the stone as shown in figure 35. The curve is called a parabola and its equation is readily obtained by algebraically eliminating t between the two equations above, to give

$$y = (v/u)x - (g/2u^2)x^2$$

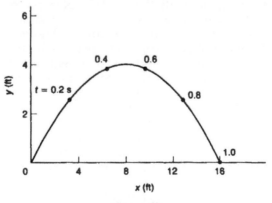

Figure 35

or, for our specific numerical example, $y = x - (x^2/16)$. In particular, the stone comes back to Earth again in exactly one second and lands some 16 feet away from where it was thrown—not exactly a Herculean effort.

Although it is perhaps not immediately apparent, the stone-throwing problem is, in reality, the 'contraction' of a two-body problem in which one of the bodies (the Earth) has been assumed to be both infinitely large and infinitely massive as compared with the stone. More general Newtonian gravitational two-body problems lead to motional paths, or orbitals, which are so-called conic sections (that is, planar slices through a cone). These sections can vary from a circle (when the plane surface is at right angles to the cone axis), through ellipses (as the plane tilts), to a parabola, and finally to hyperbolas. Since circles and parabolas are very special cases, most orbitals are one of two types: ellipses for closed orbitals (like the planets) or hyperbolas for open orbitals (like many of the comets). Interestingly, as early as 200 BC, Apollonius of Perga wrote a celebrated *Treatise on Conic Sections*, a work of eight books (of which four are extant), which systematically sets out all the properties of these curves. It was carried out as an exercise in pure mathematics (planetary motion in the days of Apollonius being organized in the Ptolemaic system with planets moving around the Earth in circles on circles). Thus it was that this mathematical study waited more than 1800 years to finally find its application in science; possibly something of a record!

From this theory of gravitation, Newton himself was able to extract an enormous amount of information concerning natural phenomena. For example, he was able to explain many motional 'laws' that had been determined to great accuracy in earlier years through the painstaking analysis of vast amounts of experimental data accumulated by previous generations of astronomers. The most famous such laws were those of the German astronomer Johannes Kepler (1571–1630) based on data collected by the Danish astronomer Tycho Brahe to whom Kepler was an assistant. In particular, his primary conclusions—namely: (a) that the planets orbit the Sun in elliptical paths with the Sun at one focus of the ellipse; (b) that the areas swept out in a planetary orbit by a straight line joining the planet to the Sun are equal for equal time intervals; and (c) that the squares of the orbital periods of the planets are proportional to the cubes of their mean distances from the Sun—were all exactly accounted for by Newton's two-body gravitational theory.

However, Newton did fully realize that the heavens could not

be separated, or compartmentalized, into isolated pairs of interacting two-body systems. The Moon, for example, is not only
gravitationally attracted to the Earth, but also simultaneously to
the Sun and the other planets. Newton tried to deal with this
problem at least for what he deemed to be the simplest (and what
was certainly a fairly common astronomical) circumstance,
namely the case in which one two-body force exerts a much larger
influence on the primary motion of concern than the others. He
developed a mathematical method which enabled him to extract,
in an approximate fashion, the forms of the small deviations from
two-body motion that were induced by the minor forces. In
particular, using this 'perturbation' theory, he was able to successfully account for many small irregularities of lunar motion
that were known to exist from prior careful observation. In fact, in
the decades that followed the publication of *Principia*, many other
previously mysterious motional irregularities of the planets were
also accounted for by the Newtonian perturbational scheme.
Included among these were the precession of the equinoxes,
tides, planetary wobbles about their axes, and eventually even
the prediction and discovery of 'new' planets, including Neptune
in the nineteenth century and Pluto in the twentieth. In truth,
probably no other law of nature has so unified such a host of
seemingly diverse natural phenomena as has Newton's gravitational law with its accompanying differential equations. So
impressive was the success of the method that the esteemed
French mathematician Pierre Simon de Laplace (1749–1827) once
boasted that given the position and velocity of every particle in
the universe he could, at least in principle, predict the future for
the rest of time. The philosophical conclusion, that human
behavior was completely determined, and that therefore free will
was an illusion, was (not surprisingly) very troubling to many —
but it certainly did seem to be an undeniable consequence of the
equations involved.

   And yet, lurking within this success story (and 'tucked away' in
the 'in principle' insert of the Laplacian claim) lay the spectre of a
profound and diabolical frustration. It was the fact that even the
general Newtonian three-body problem (let alone the far more
complex motions envisioned by Laplace) gave rise to differential
equations for which no *exact* solution existed in terms of known
mathematical functions. On the other hand, it *was* clear that *any*
set of Newtonian differential equations was undoubtedly deterministic. By this we mean that, if you fixed the starting positions
and velocities of the bodies concerned at any particular time, then
the subsequent motion (whatever the forces involved and what-

ever the precise mathematical description) must be uniquely defined. For the case of Newton's gravitationally generated two-body elliptical orbits the answer was simple. However, its essential simplicity lay not primarily in the fact that the orbit could be expressed in a known functional form, but rather because the motion repeated itself endlessly. But what of other possible circumstances? Might some of the Newtonian sets of equations possibly generate orbits that *never* repeat? And if so, what would this imply?

The first person to ponder this aspect of the problem in depth was the French mathematician Jules Henri Poincaré (1854– 1912). Since even the three-body problem in its full Newtonian form was of unapproachable mathematical difficulty, he decided first to focus on what should be a simpler limiting case—namely that for which one of the three bodies is so tiny that its motion does not perceptibly affect that of the other two. We may think, perhaps, of a grain-of-sand-sized meteor wandering into the spatial vicinity of the Earth and the Moon. Within this model, the Earth and the Moon are allowed to maintain their elliptical orbits exactly. But what of the trajectory of the tiny meteor? How simple or complicated is its orbital if the problem, as posed, is solved exactly?

Much to his disappointment, Poincaré soon found that even this over-simplified version of the Newtonian three-body problem defied exact solution in terms of known mathematical functions. Therefore, lowering his sights a little, he concentrated on searching for some general characteristic of the motion that might at least shed some light on the nature of the orbital in question. Specifically, he devised a method by which he could test whether the orbital, whatever it might be, ever retraced itself. In essence, he set up a plane and checked the numerically calculated orbital position every time it crossed the plane. In this manner he was able to conclude that, in general, the grid of intersection points (coupled with the information concerning the direction of the orbital motion on each passage) would never repeat itself. From a purely computational point of view it was more than a little disappointing. 'All of this is a bit complicated and counterintuitive,' he scribbled in his notes, together with other such discouraging statements as 'the intersections form a kind of net, or infinitely tight mesh . . . one is so struck by the complexity of the figure that I shall make no attempt to draw it.'

Discouraging though this may have been, Poincaré's finding did in fact carry with it the seed of an even deeper trouble which was not then recognized. Orbitals of this kind are not only

infinitely complicated, they possess an unimaginable sensitivity
to their starting conditions—the essence of what is now known as
chaos. By this we mean that even the tiniest change in either the
position or velocity vector at starting time $t = 0$ eventually
accumulates to alter the orbital, after sufficient time has elapsed,
by extremely large amounts. More precisely, after a sufficient
time, two orbitals deduced numerically from the same set of
equations and from starting conditions that are arbitrarily close,
will differ to the extent that any correlation is completely unrecognizable. However, in the days before the advent of the
electronic computer, the full details of these effects were beyond
numerical calculation and the true essence of the meaning of
chaos—and, in particular, of its difference from ideas of
randomness—could not be appreciated.

Mathematically, the problem confronting us is that of solving
what are called 'non-linear' equations, and the fact that very few
such equations can be exactly solved by analytical methods. A
linear equation is one for which the sum (or addition) of any two
separate solutions also satisfies the equation. As a subset of all
equations, linear ones are by far the simplest to attack. Perhaps
the most famous of all linear differential equations is that for
small oscillations (or simple harmonic motion), which may be
written

$$d^2x/dt^2 = -n^2x$$

in which n is a constant with the dimensions of reciprocal time (or
frequency). In various guises this equation has dominated much
of twentieth century solid state physics (see chapter 4). One
solution is readily verified to be $x = \sin(nt)$, while a second is
equally easily seen to be $x = \cos(nt)$. The linearity of the equation
therefore dictates that

$$x = \sin(nt) + \cos(nt)$$

should also be a solution—and, again, it is not difficult to check
out that such is the case. So thoroughly has the equation for
simple harmonic motion been studied in both classical and
quantum limits (where quantum theory is that theory which
gradually takes over from Newtonian mechanics at atomic
dimensions) that virtually all problems which could possibly be
'forced' into harmonic form (by neglecting or approximating
difficult and hopefully small non-linear terms), have been presented in this form.

It is exactly in this manner that such concepts as 'phonons' (the
harmonic vibrations of the collective motion of atoms about their

lowest-energy ordered lattice positions in a crystal) and 'magnons' (the harmonic angular oscillations of tiny atomic magnets about their lowest-energy fully aligned state) arise in solid state physics, see chapter 4. The validity and usefulness of concepts such as these, and of the approximations made in order to generate them, naturally rest upon testing the theoretical predictions of measurable properties to which they lead against experiment. In most cases, happily for the solid state physicist, the results have been quite satisfactory. Mathematically, in the solid state context, the relevant equations typically involve the interactions between about 10^{23} atoms. Faced with that sort of complexity, it is not at all surprising that mathematical rigor is often asked to take a back seat. But, as we shall see, it is not only in the complex context of the modern quantum theory of solids that theoretical physicists 'cheat' in order to appear to 'solve' tricky non-linear almost-harmonic equations.

Surprisingly, a very common example can be found in nearly all High School and College physics text books in the discussion of the very familiar classical motion of a simple pendulum. Undisturbed the pendulum hangs in a vertical position. If held to one side by a small angle and then released it begins an oscillatory motion. A discussion of the physics of the problem shows that, during the motion, the restoring force on the pendulum, which is the force controlling the motion, is almost (but not exactly) proportional to the angle of displacement from the vertical. The smaller the angular displacement, the smaller is the deviation from exact proportionality (for which limit the old faithful equation of harmonic motion arises once more). It follows that for the oscillations of a simple pendulum with any non-zero amplitude there are always small terms present in the equation of motion that make it 'hard' to solve. It is the philosophy of the text book to neglect them in the hope that, at least for oscillations deemed to be small in some sense, the harmonic solutions are 'valid'.

Now this is the equation of motion of a single classical oscillator. Small and difficult terms should surely not be shrugged off for such a seminal case with the same abandon used by the solid state physicist when confronted with some 10^{23} interacting oscillators. And yet, to my knowledge, this is exactly what is done in almost all, if not all, elementary text books. The important question that is being begged is whether an exact solution to an approximate equation is necessarily a good approximate solution to the exact equation. From our story concerning the Newtonian three-body problem, and of the existence of chaotic orbitals, we suspect that the general answer is no—and such is indeed the case. For the particular case of the simple pendulum with small oscillations, however, the answer (at least when compared with experiment) seems to be yes, and it is for this reason that the matter is happily glossed over in the text books. A rigorous mathematical proof of the 'yes' answer for the case of the simple pendulum did not, in fact, appear until the latter part of the nineteenth century, long after many generations of pupils had happily assumed it out of deference to their unquestioning text books. We therefore learn that small non-linearities do not necessarily give rise to demonic properties. Unfortunately, however, in many cases they do, and to a degree that was not fully recognized until electronic computation entered the picture.

In this context, our story now moves on to one Edward Lorenz, of the Massachusetts Institute of Technology, and to the early part of the 1960s. Lorenz was a meteorologist and was, at that time, concerned with improving the accuracy of weather forecasting by using the latest electronic computers to produce ever more accurate numerical solutions to his model equations. On this particular occasion he was considering the simultaneous solution of a set of 13 differential equations that purported to represent weather patterns in the atmosphere. After following their time dependence for several hours of computer time, he stopped the 'run' and took a day or so off to consider his results. Some days later he decided to extend the same run to longer time spans, so he restarted the computing procedure but, just to check that everything was running satisfactorily, he didn't restart at the end of the previous run, but at some modestly earlier point (so he could verify that the program was indeed operating correctly by regenerating the latter part of his earlier run). However, for his new starting numbers he punched in values to only three significant figures whereas the computer, of course, was working to a much greater accuracy. What he observed confounded him. As the new numbers began to be printed out, they followed the earlier results (to three significant figures) for a while, but then the pattern began to change until soon it bore no resemblance whatsoever to the earlier run. His first suspicion was of a computer malfunction, but eventually the truth began to dawn as he probed the curious phenomenon in more detail. It appeared that, for his equations, even the tiniest changes in starting numbers eventually produced enormous changes in output if the run was allowed to proceed far enough.

He decided to probe the phenomenon at a more basic level by reducing his set of equations to their absolute essentials. He found that the effect persisted, even when the atmospheric model was reduced to one of simple convection. Mathematically, this 'bare bones' model involved only three variables (say x, y, z). They were related to their time derivatives by equations of the form

$$dx/dt = Ax + By$$

$$dy/dt = Cx + Dy + Exz$$

$$dz/dt = Fz + Gxy.$$

Now the letters A to G were, for any particular example, simply numbers (like $A = -10$, $B = 7$, $C = 25$, and so on) which were decided by the physics of the problem. A glance at the equations will reveal that they are linear if E and G are both zero. For this case it was possible for an accomplished mathematician like Lorenz to solve them exactly (that is, in analytic form) without the need for any computer assistance. Unfortunately, the essence of convection requires the presence of non-zero values for both E and G, and hence the convection problem was inherently non-linear and required numerical 'solution'; solution, that is, to the accuracy of which the computer was capable. Lorenz therefore reprogrammed his computer once more in order to examine the new problem.

Starting from solutions where all the terms on the right-hand side are zero (that is, for which x, y, and z remain constant) he changed one of the constants A to G just a little bit, and then let the computer work out the time-dependent consequences of the change. Since the change in the equation of motion from the steady state was extremely small, the naive expectation might be that the solutions x, y, z, as functions of time t would drift only very slowly away from the steady state; and for a while this appeared to be true. But then the solutions began to wobble periodically about the steady state in a manner that became increasingly erratic, and finally everything went crazy, with solutions swinging wildly through vast ranges with no patterns any longer recognizable. It was clear that, even for this 'bare bones' set of non-linear equations, the tiniest changes in the equation parameters eventually produced wild fluctuations in the solutions.

This was the shattering of a weatherman's dream. 'I knew right then,' Lorenz said, 'that if the real atmosphere behaved like this, then long-range forecasting of weather was simply not possible'. In the broad context of global weather it meant, presumably, that an almost inconsequential change in starting conditions (say a fraction of a degree change in temperature at some specific location) could result in the difference between a day of sunshine or one of torrential rain at that, or some other, location a month or two later. In the absence of *perfect* numerical precision concerning every relevant quantity, even an approximate degree of prediction into the far future would be impossible. Lorenz called this the 'butterfly effect', suggesting that even the flapping of a single butterfly's wings could produce changes that, sufficiently far ahead in time, might give rise to major atmospheric disturbances. From a computational point of view the conclusion was equally dramatic. Since no computer can calculate with absolute precision, the mere 'rounding-off' of numbers by the computer, no matter to how many decimal places, itself introduces sufficient inaccuracy to make any calculation in this context eventually worthless in the long run.

Lorenz noted something else about the numerical solutions to the above set of three convection equations. It was that, even when finite non-linear terms involving parameters E and G were present, not all values for the other parameters gave rise to chaotic solutions. As one progressed gradually along a sequence of values for one of the parameters, while holding the others constant, the solutions appeared to pass from normal (that is, non-chaotic) to chaotic. Thus, although the parameter values that were most relevant for describing real convection physics did produce chaotic solutions of the kind described above, other sets of non-linear equations of exactly this same differential form did not.

Exactly how does orderliness degenerate into chaos as a function of some slowly varying parameter? For many years the explanation credited to the Soviet theoretical physicist Lev Landau (1908–1968) held sway. It contended that chaos results when a gradually increasing number of independent oscillations enters the motion, one at a time, as the order-to-chaotic transition regime is approached. In this way, although each separate oscillation may be simple, the increasingly complex combined variations finally render the resultant motion impossible to predict. But such is not the case, as we shall see. Chaos can, in fact, arise in the most orderly of fashions, and in systems far simpler than the convection model set out above; ones that do not even require differential equations for their description.

For a full understanding of this fact it proved necessary to await the appearance of just such a 'simple' model. And when it appeared, it proved to be so simple that even a pocket calculator was sufficient to probe many of its mysteries. Mathematicians refer to it as the problem of logistic mapping, but it is much more readily understood when set out in less abstract terms. The example usually cited is a model for the study of the population fluctuations of seasonally breeding insects. Suppose, for example, that a particular environment is able to support an equilibrium (that is, stable) population of, say, 500 000 insects. Implicit in this statement is the supposition that, if the population in any particular breeding season were less than this, then the environment would favor an increase in numbers. Also, if the insect numbers were more than 500 000, then the environment would be insufficient to support this over-population and this would lead to a corresponding decrease. A particularly simple non-linear equation which can model this situation from breeding season to breeding season may be written in the form

$$x_{n+1} = 2x_n(1 - x_n)$$

where $10^6 x_n$ is the insect population in season n (that is, x is measured in units of one million). Given the insect population in any one season n, the equation then allows us to calculate the model's prediction for the population in the following season, namely season $n + 1$.

The fact that 500 000 insects (that is, $x_n = 0.5$) is a stable population for this 'generating' equation is easily verified since, if we put $x_1 = 0.5$ into the above equation, then we immediately generate values $x_n = 0.5$ for all values of n greater than one. But what if the starting population value takes some value different from 0.5? Try $x_1 = 0.8$, for example, and take out your pocket calculator. It doesn't take long to generate the pertinent sequence of values (that is, x_n with $n = 1, 2, 3, 4, \ldots$) as follows:

$$0.8, 0.32, 0.4352, 0.4916, 0.499\,86, 0.499\,999, \ldots$$

which establishes that, by the sixth 'season' we have returned to the stable population. In fact, as you may easily verify, it is possible to start with any number between 0 and 1 for x_1 and observe a convergence to the stable limit of $x = 0.5$. This stable value is known, in circumstances of this kind, as a 'fixed point' and, since all starting numbers within the stated 'band of values' between zero and unity inexorably lead to it as an endpoint, it is also referred to as a *stable* fixed point, or as an 'attractor'. This is important because the above generating equation has another fixed point that may have escaped your notice. It is $x = 0$. However, this fixed point is unstable in the sense that no matter how small a non-zero value you choose for x_1, the sequence of subsequent values $x_2, x_3, x_4, \ldots$ now always moves steadily away

On the Shoulders of Giants

from $x = 0$ towards the attractor at $x = 0.5$. But, in spite of its fascination, these sequences are far from chaotic. On the contrary, they each progress in the most orderly of fashions as they 'zero-in' on the stable fixed point.

But now let us suppose that our stable insect population is somewhat larger than 500 000; implying that we have improved the environment in some manner, the details of which are of no concern to us. Let us suppose that the new stable population is 687 500, corresponding to an x-value of 0.6875. There is nothing magic about this number, and I have chosen it solely because it is significantly larger than $x = 0.5$ (which is the direction I wish to investigate) but retains a generating equation that is still numerically very simple, namely

$$x_{n+1} = 3.2x_n(1 - x_n).$$

Inserting $x_1 = 0.6875$ into this equation, we readily verify that $x = 0.6875$ is, indeed, a fixed point. But is it an attractor? If you check it out using your calculator (starting from some number arbitrarily chosen between 0 and 1) then you will soon deduce that it is not. A rather strange phenomenon takes place. The sequence of seasonal populations neither regresses from, nor converges towards, the known fixed point $x = 0.6875$. Rather it settles down to a repeating sequence that oscillates back and forth between two values 'a' and 'b' in the manner *abababababab* ... where

$$a = 0.799\,455\,490\ldots$$
$$b = 0.513\,044\,510\ldots$$

to the nine-digit accuracy of my calculator. These two values 'bracket' the fixed point of 0.6875 but, perhaps somewhat surprisingly, not in a symmetric fashion since

$$(a + b)/2 = 0.656\,25.$$

In spite of this unexpected behavior, the physical implication in terms of insect populations remains quite meaningful in the sense that an under-population always increases in the following season while an over-population decreases.

Inspired by this rather curious behavior we proceed to increase the equilibrium (fixed-point) insect population a little further; this time to $x = 5/7$, for which the generating equation is

$$x_{n+1} = 3.5x_n(1 - x_n).$$

Although this now strictly implies a stable population involving a fraction of an insect ($10^6x = 714\,285.714\ldots$) you will, I hope, overlook this quirk of insect reproduction by mathematical modeling and focus once more on the numerical behavior of the generating equation. Starting again with any number between 0 and 1 for x_1, we now observe an even more curious pattern. The sequence of numbers x_n now always eventually settles down to the repetition *abcdabcdabcd* ... where

$$a = 0.382\,819\,683\,\ldots$$

$$b = 0.826\,940\,707\,\ldots$$

$$c = 0.500\,884\,210\,\ldots$$

$$d = 0.874\,997\,264\,\ldots.$$

We have now progressed to a cycling pattern of four numbers, and once again they do not position themselves symmetrically about the fixed point of $x = 0.714\,285\,\ldots.$

As the numerical constant A in the generating equation

$$x_{n+1} = Ax_n(1 - x_n)$$

is increased further, the character of the stable limiting number pattern changes further, first to an '8-cycle', then to a '16-cycle', then a '32-cycle', and so on until the number of repeating numbers in the cycling pattern diverges as 2^n, with n progressing steadily through the infinitude of counting (or 'natural') numbers 1, 2, 3, 4, 5, 6, ... And then, finally, beyond a limiting value for A (which is close to 3.57) the length of the 'repeat cycle' becomes truly infinite or, in other words, no repetition remains and the number sequence becomes completely chaotic. In this chaotic regime, two starting populations chosen arbitrarily close in value eventually lead to corresponding sequences x_n that possess no correlation whatsoever.

We therefore possess, in this simplest of generating equations, a complete progression of behavior from orderliness, through an infinite sequence of cycle-doublings, to chaos. Let us now look a little closer at the exact manner in which this transformation to chaos takes place as regards the numerical details. The first person to think seriously about this aspect of the problem was Mitchell J Feigenbaum, a physicist working at the Los Alamos laboratory, in 1975. His first step was to compute the exact sequence of values $A = A_n$ at which the doublings to an n-cycle (that is, with 2^n numbers in the repeat cycle) takes place. The actual values themselves are not important, although they rapidly get closer and closer together as the point $A = A_x$ (for which $n \to \infty$) is approached. What Feigenbaum did find to be of

Mitchell J Feigenbaum, 1944–. Toyota Professor of Physics at The Rockefeller University. Courtesy of M J Feigenbaum.

the greatest interest was the fact that the decreasing gaps between successive A-numbers settled down to a most regular behavior as the chaotic limit was approached, each one being a little over four and a half times the next one. More precisely, they approached the limiting form

$$A_{n+1} - A_n = d(A_{n+2} - A_{n+1})$$

where

$$d = 4.669\,201\,609 \ldots$$

is an irrational number, now known as the Feigenbaum constant.

Feigenbaum, being a physicist, saw in this relationship the essence of a previously recognized phenomenon known as 'scaling', or the recurrence of features with exactly the same structure on ever-smaller scales (we shall meet this again in discussing the concept of 'fractals' in chapter 11). Scaling had first become well known to solid state physicists in the early 1970s in connection with a theoretical understanding of certain structural instabilities in crystals (known as 'order–disorder phase transitions'). These are the sharp changes of state from order to disorder of the kind seen in such physical phenomena as the disappearance of magnetic order (ferromagnetism), electric order (ferroelectricity), or atomic component order (ordered alloys) with increasing temperature in some solids. Many other examples from the field of solid state physics could be given, but the important numerical finding (first from experimental observation and then from theoretical modeling) was the fact that whole classes of seemingly different transitions tended to scale mathematically on approach to their instabilities in *exactly* the same manner—a property known as 'universality'. In this fashion, the vast host of possible phase transition types could be lumped together into a relatively small number of different 'universality classes'. With this fact in the back of his mind, Feigenbaum was therefore led to examine other kinds of simple non-linear generating equations. He chose next the trigonometric form

$$x_{n+1} = A \sin (x_n)$$

and again found a developing scaling behavior on approach to chaos. Anxiously, he once more computed the scaling constant d to great accuracy and, lo and behold, that same hitherto unheralded number beginning 4.669 201 609 showed up again.

Probing now more widely, Feigenbaum discovered that all generating equations of the form $x_{n+1} = f(x_n)$ for which the function f, when plotted in the normal Cartesian form $y = f(x)$, had the shape of a rounded hump (as do $y = Ax(1 - x)$ and $y = A \sin(x)$), exhibited this very same scaling-constant d-value. For functions of a more complicated shape (involving, say, multiple humps, or humps with flattened or pointed tops) scaling still occurred, but the numerical value for d was different. However, each d-value remained constant within its own 'class' of generating functions. Thus, the infinite number of possible generating equations got lumped together into universality classes within each of which d was constant. The analogy with the universality classes of solid state physics was striking.

Although Feigenbaum was only studying simple numerical functions, he believed that this general behavior expressed a natural physical law about systems on the point of transition from ordered to chaotic motion (or 'turbulence'). And since the simplest non-linearity giving rise to such an event was represented by a smooth single-humped function, he proposed that the period-doubling route involving the scaling number $d = 4.669 \ldots$ should manifest itself in real physical contexts. Moreover, there were other numbers that he could predict as well. For example, if

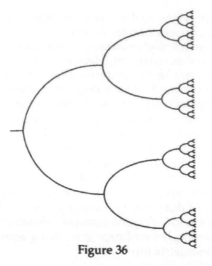

Figure 36

you plot the period-doubling events graphically, as shown schematically in figure 36, each branching event has not only a horizontal projection (or 'shadow') that measures d, but also a vertical one that measures how fast the branches open out. This latter scaling defines a second universality constant which, for the case under present discussion, takes a value of 2.502 907 875 . . . .

In spite of the numerically very specific nature of the predictions, physicists in general (and editors of science journals in particular) were exceedingly skeptical. Years later, Feigenbaum kept in a desk drawer of his office the large number of rejection letters from highly reputable journals that had refused to publish his work. They all lived to rue the day, for Feigenbaum's work proved to be a turning point in physics; a spark that ignited what was quickly to become a blaze of activity in a new science that was to cross the boundaries of so many previously isolated and esoteric scientific disciplines.

Initially, however, as far as most physicists could tell, the interesting computational regularities that Feigenbaum had found, although they did look somewhat like the onset of chaos in a mathematical sense, appeared to have no obvious connection to real physical systems. The known physical systems of the time involved vastly more complicated phenomena than anything considered by Feigenbaum. In particular, even in their most simplistic representations, they usually involved several differential equations, all of the non-linear kind (that defy exact solution). Feigenbaum's belief, on the other hand, was that, no matter what the complexity of a real physical situation, any underlying scaling properties that it possesses on approach to a chaotic instability should be the same as those of even the simplest equation within its universality class. He also believed and hoped that in many cases this behavior would be just that of his one-parameter insect-population example.

The first actual experimental evidence that the period-doubling route to chaos was more than a numerical quirk of a simplistic non-linear generating equation was obtained in France by studying the onset of turbulence in liquid helium at temperatures close to absolute zero. Now liquid helium exists only at temperatures close to absolute zero, and requires rather special methods to produce it. Its advantage, however, is that it can be obtained in a very pure state, and experiments involving its properties can be made with extreme precision since, near the absolute zero of temperature, virtually all the thermal agitation of atoms has ceased. Although this last statement would be true for any material, only helium remains a liquid (and is hence subject to fluid motion) down to these extremely low temperatures.

In 1977, Albert Libchaber, of the Ecole Normale Supérieure in Paris, decided to study the onset of turbulent motion in liquid helium. To do this he manufactured a tiny chamber (small enough to be carried around in a matchbox), and into an even smaller rectangular cell inside the chamber he fed liquid helium chilled to within four degrees (Celsius) of absolute zero. Below the cell sat a bottom plate of pure copper, and above the cell was a plate of sapphire crystal (materials chosen for their heat conducting properties). There were also tiny heating coils and elaborate mechanisms for controlling temperature and for removing any stray vibrational perturbations. Temperature differences of only a few thousandths of a degree between the top and bottom plates were sufficient to produce a convectional motion in the helium. In fact the whole point in making the apparatus so small was that a larger cell would require even smaller temperature differences to begin the motion to be studied, so small that they could not be adequately controlled or reliably measured.

The initial thermally induced motion of the helium was periodic, creating a pattern somewhat like a row of sausage rolls (see figure 37). The physics of this kind of periodic motion was already well known and well understood. In order to see what happened as the temperature difference was slowly increased, Libchaber embedded tiny temperature probes in the upper sapphire plate, recording their outputs continuously by a pen plotter. As the

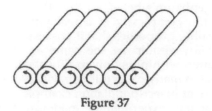

Figure 37

bottom plate was heated the rolls began to develop wobbles, although the rotational periodicity of the convection was maintained. But then, quite suddenly, a new oscillation appeared with exactly twice the previous period of oscillation. And just above this temperature difference, oscillations at 4, 8, and 16 times the original period could clearly be seen to appear in sequence. Beyond that the details rapidly became scrambled and the chaotic motion referred to as turbulence set in. In spite of the limitations in numerical precision, the experiment clearly established a single fact of extreme importance. There was no doubt whatsoever that the approach to chaos in this real physical system of convection transforming to turbulence occurred via a period-doubling cascade.

At the time of the initial experiment, Libchaber was unaware of Feigenbaum's numerical predictions so that no search for scaling in the period-doubling was made. By 1979, however, the contact had been made and Libchaber took a second look at his data. What was the scaling ratio for his experimental period-doublings? Firstly, was it even constant? Within experimental accuracy the answer to that question appeared to be yes, so that some scaling behavior was taking place. But was the scaling number or ratio anywhere near Feigenbaum's 4.669? Again the answer seemed to be a qualified yes, but greater accuracy in the experimentation was clearly needed. With excitement now growing, many other experiments for studying the onset of chaos were rapidly devised over the next year or two, not only in the context of fluid turbulence, but also in electronics, optics, and even chemical reactions and biology. The results proved to be very gratifying, with Feigenbaum's predictions confirmed to a greater and greater accuracy and in an increasingly wide context of instabilities.

It all seemed a bit like magic. Somewhere within each complicated set of non-linear differential equations describing the instability in question, even though they were all quite different in detail for each experimental context, was evidently buried that single-humped non-linearity that Feigenbaum had probed numerically. It was as if nature had a penchant for simple one-humped one-parameter dynamics. However, as experimental activity now exploded on the subject of the onset of chaos, examples were also found of some of the other (non-single-hump) routes, and these also proved to exhibit scaling and universality in accord with theoretical expectations. These discoveries also initiated a flourish of activity in the field of computer calculation. As computers got faster and faster and were made with larger and larger storage memories, they were finally able to numerically attack realistic models containing complicated sets of non-linear differential equations. In this context it was then possible to carry out accurate computer 'experimentation' of the onset of chaos, and to do so faster and with greater precision than was feasible for real experimentation. In this manner the period-doubling avalanche route to chaos was 'observed' and precisely probed in many a computer 'experiment', and Feigenbaum's scaling constants were verified to increasingly greater precision.

For most people, however, no amount of numerically complex computer 'experimentation' can replace the fascination of observing the onset of chaotic motion in objects of everyday familiarity. In this category, one of the most common examples is that of the dripping tap. Presumably almost everyone has noticed that the dynamic behavior of a dripping tap varies rather dramatically as the rate of flow is gradually increased. A detailed description of the physics of the event is unfortunately rather complex, with water drops hanging from the tap and gradually growing and changing shape before finally undergoing a complicated instability involving the production of a narrow neck and subsequent 'break-off'. The problem involves concepts like surface tension and viscosity, and the accompanying mathematics at this level of sophistication is frightening. But if you forget altogether about these complications and just concentrate on the series of numbers that measure the intervals of time between drips (recorded, perhaps, by a light beam that is intermittently broken by the drips as they fall), then all the beauty of the transition from order to chaos is present in these numbers alone. For a slow water rate the drips are 'regular' (that is, taking place at constant intervals of time), and the numerical pattern of time intervals is therefore a sequence of identical numbers like $a, a, a, a, a, a, a, a, \ldots$ As the rate is slowly increased, a point is reached for which the drops suddenly begin to fall in pairs; that is, with a number pattern looking like $a, b, a, b, a, b, a, b, \ldots$. And then, at a still larger water rate, comes a pattern of four numbers $a, b, c, d, a, b, c, d, \ldots$, and then eight, and so on in a 'period-doubling' cascade to chaos. Indeed, if any of the taps around you has a deteriorating washer, this actual process may be taking place in your own home at this very moment. Once again, the mathematical equations that accurately describe the complete physics of the problem are fearfully complicated in detail. But, regardless of this fact, it must be true that they also contain, buried somewhere within their elegant complexity, the same old single-hump non-linearity of Feigenbaum's original insect-population-generating equation.

Impressive though all of these 'discoveries' are, they still tell us little or nothing about the chaotic state itself. A lot has now become clear concerning the path by which ordered motion approaches chaos *from the ordered side*, but what about the motion on the chaotic side? Is it the same as random motion—or is it much more than that? Was Poincaré, all those years ago, prescient when he threw up his hands in despair on consideration of the gravitational three-body problem? What happens to Feigenbaum's numbers when the truly chaotic regime is reached? In order to probe this last question it is merely necessary to go back to the original insect-population equation

$$x_{n+1} = Ax_n(1 - x_n)$$

and examine the number pattern for some value of A (say $A = 4$) that is within the chaotic regime. If you do this with your pocket calculator, starting with any number between 0 and 1, you will find that the numbers generated do have a distinctly random-like appearance. However, if you program a personal computer to do the job, and proceed to a few thousand places in the number sequence x_n, weird and wonderful apparitions await. In the midst of the random-like complexity, quasistable cycles of numbers suddenly appear with cycles of odd period, like 3 or 7, before another period-doubling sequence destroys them. Quite evidently there are number patterns of many kinds hidden in the sequence—not something that one would expect for a truly random number sequence at all.

One of the best ways of 'seeing' these hidden patterns is to draw a picture or, since the picture is very detailed and fine textured, get a computer-graphics package to do it for you. The idea is to take the starting number as a measure of distance along an *x*-axis in a graphical *xy*-plot and the first-generated number as its counterpart *y*-coordinate, then to 'move on' to take the second and third generated numbers as the next (*x*, *y*) pair, and so on. In practice it is best to let the numbers run a little before starting the plot in order to enable the chaos to 'settle down'; that is to remove any bias inserted by the initial choice of starting number.

If the generated numbers were truly random, then the graph paper would gradually become covered with a random-like array of dots covering the entire plot (within the bounds set by the problem). But this is not what happens at all. In actuality, a very precise pattern builds up, albeit an extremely complicated one, with whispy loops clinging to the backs of other loops. A very definite structure is seen to be unmistakably present. Moreover, this structure is an attractor in the sense that, at least within a certain range of starting numbers, it persists as the stable chaotic signature of the particular brand of chaos under examination. It is, in fact, every bit as much of an attractor as was the simple point $x = 0.5$ in our first examination of the insect-population problem. And yet, because it is so immensely complicated in its detailed structure, mathematicians (or chaos researchers if you prefer) refer to it as a 'strange attractor'. These strange attractors are not necessarily confined pictorially to a two-dimensional (x, y) description. By taking the generated numbers three at a time, rather than two at a time, and relating them to three Cartesian coordinates (x, y, z), three-dimensional manifestations of these apparitions are just as readily constructed. In fact, the procedure can be generalized to four, five, six, or more dimensions, although the task of 'viewing' these higher-dimensional patterns obviously presents more of a technical challenge.

What we have found is evidence that there is, buried deep within any chaotic behavior, hidden patterns of order. It still remains true that the tiniest changes made in the 'coordinates' at any time whatsoever eventually alters the subsequent sequence of points to the degree that it becomes unrecognizable; indeed, this is the essence of chaos. This change, however, takes place in the sense only of the sequence in which the points appear on the attractor. All points still remain on the attractor, which remains the stable configuration. We have, in essence, complete local unpredictability, but complete global stability. In the context of weather, this might mean that although day-to-day forecasting far into the future is futile, the locally erratic behavior precluding it, it still remains true that within the larger 'global' picture the range of temperatures in summer will always be higher than those in winter. Buried beneath the local erratic behavior there is a longer-period stability although, to my knowledge, no specific strange attractor for weather has yet been determined.

The concept of attractors (strange or otherwise) is equally

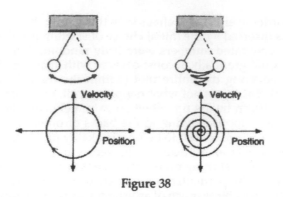

Figure 38

viable in cases of continuous motion, or dynamics if you prefer. In cases of this kind, mathematicians like to make use of the concept of 'phase space'. Consider, for example, a simple pendulum (oscillating in one dimension). All that are needed to determine its motion are two variables, position and velocity (or, equivalently, momentum), which may therefore be plotted at an instant of time as a point in a plane whose axes measure these two quantities. This plane is then referred to as the phase space of the pendulum. Within this phase space, frictionless oscillations are easily seen to be represented by a circle (see figure 38). This circle is not an attractor since, if you perturb the pendulum motion by a small amount, you also permanently perturb the phase space circle that defines the motion. However, in the presence of friction, an attractor can be found. In this case the oscillations damp out with time and the phase space orbit becomes a spiral (figure 38) which eventually 'strangles' the origin (0, 0) of the plot. This singular point is now the attractor of the motion since it remains the common endpoint regardless of the starting conditions.

For an orbiting three-body problem in three dimensions it takes no less than 18 numbers to define a point in phase space; nine for the positional coordinates (x_1, y_1, z_1), (x_2, y_2, z_2) and (x_3, y_3, z_3), and another nine for the velocity coordinates. The phase space for this problem therefore is one of 18 dimensions. Mathematically there is no difficulty in defining Cartesian spaces in any number of dimensions, even up to numbers of the order 10^{23} for macroscopic solid bodies, but within these spaces (no matter how grandiose) the state of the system at any one instant of time is still denoted by a single point. The 'motion' of the entire system is therefore completely defined by the motion of a single point moving through its phase space. For periodic systems this point will always move in a manner which, no matter how complicated its pattern, eventually repeats itself. For a chaotic system, on the other hand, the trace of the point never repeats. But neither does it fill the phase space completely — rather it defines another swirly and tortuously complex phantom-like form — the strange attractor. Although significantly different initial conditions can evolve towards different attractors, all the points within a so-called 'basin of attraction' always evolve towards the same attractor, and the lower the dimension of the attractor, the 'simpler' is the chaos. But to fully appreciate even the simplest of strange attractors, it is really essential to have the assistance of a computer-graphics package because these curious objects are, in general, of a fractal nature (see chapter 11). This was Poincaré's dilemma. In his pre-computer era he could do little more than throw up his hands in despair, although it is quite clear from his essays that he recognized, far more than his contemporaries appreciated, many of the fundamental properties underlying the description of what we now refer to as chaotic motion.

Now that the basic mathematical framework necessary for detecting chaos is well understood, chaotic behavior is seemingly being sought in every imaginable context from the Gulf Stream to the Stock Market, from heart attacks to measles epidemics, and from the tumbling motion of Hyperion (a potato-shaped moon of Saturn) to the orbit of the planet Pluto. Success has been achieved in some cases (Pluto's orbit does appear to be chaotic) while a serious study is still pursued in many others (the electrical activity of the heart and brain *may* be chaotic). However, all these examples are from the macroscopic world. Very recently, a prime focus of attention has centered on the possibility of chaos taking place at the atomic level.

The first experiment performed to detect atomic chaos involved two charged atomic particles (actually charged atoms of barium) that were held confined in a magnetic 'trap'. Perhaps it is simplest to picture the charged atoms (or ions) as balls, and the trap as a double saucer (figure 39) in which one ball can sit in equilibrium close to the bottom of each indentation of the saucer. We say 'close to the bottom' and not actually 'at the bottom' because of the presence of the repulsive electrical force between the two positively charged ions which attempts to prevent them from coming together. Suppose now that the shape of the double saucer is progressively changed in the manner depicted in the figure. A competition develops between the gravitation-like forces (physicists say 'potential energy') of the saucer, which

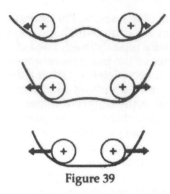

Figure 39

attempt to push the atoms closer together, and an increasing electrical force that tries to keep them apart. Working with the proper equations of motion for the system, computer solutions suggest that a critical situation is eventually reached for which the motion changes from one of localized (thermally driven) oscillations about the two separate equilibrium positions to one of chaos. Using laser light scattering off the two atoms, physicists were able to directly observe the experimental atomic system and were able to verify quantitatively the theoretical predictions.

But small though barium atoms are, they still do not quite take us into the realm of smallness for which the laws of quantum mechanics (see chapter 12) take over from the classical laws of Newton. In order to probe the effect of chaos in the true quantum regime it is necessary to focus upon the motion of electrons. The simplest example concerns the properties of the lightest of all atoms, hydrogen, which consists of a single electron 'orbiting' a more massive protonic nucleus, the force of attraction that binds them together being the electrostatic force between particles of opposite charge. The question to be asked is whether chaos, that special kind of disorder which crops up so frequently in the macroscopic world, also plays a role in the world of quantum physics? The added difficulty confronting the investigator in the latter case is the fact that a quantum description of the electrons in the hydrogen atom (or any other atom for that matter) already contains within it, by its very nature, both motional and positional uncertainty.

The electronic 'motion' of an electron about a nucleus cannot be cast, in quantum theory, in terms of classical concepts like orbits. The position of the electron is described by a wave pattern, which takes the form of a cloud-like smear, hovering near the proton, and this pattern does not change with time. This smear (called the wave function) is related mathematically to the probability of finding the electron at any particular point. In other words the electron cannot be pinpointed in space, and only questions related to probabilities have experimental relevance in quantum mechanics. However, the electron can exist in any of an infinite number of discrete states (called energy levels), each of which has its own wave pattern associated with it. The lowest energy level of all is called the ground state of the system, from which level the atom can be 'excited' by injecting energy in the form of heat or electromagnetic radiation, for example.

In the light of this positional uncertainty of the electron, which is forced upon the system by the very nature of quantum theory, researchers have had a difficult time deciding what quantum chaos should be, let alone how to look for it. One renowned physicist has been quoted as saying 'the very term "quantum chaos" gives people the shivers'. Evidently it is to be sought as some randomness existing over and above that inherent in the nature of quantum mechanics. Where would be the best place to look? Well, since there is no sharp boundary between the quantum and classical regimes, one can ask what happens to the quantum description of a classically chaotic system as it smoothly passes over from the classical to the quantum domain.

The hydrogen atom approaches the classical domain at energies close to those that will remove the electron completely from the vicinity of the proton. As mentioned, the atom is not free to assume any arbitrary energy. It can take only discrete (or 'quantized') values corresponding to the 'allowed' energy levels. At low energies these levels are spread relatively far apart. But when the energy of the atom is increased, the atom grows bigger (as the electron spreads out farther from the proton) and the energy levels get closer together. Eventually, at energies just below that sufficient to cause the electron to 'escape', the energy levels become so dense as to be almost continuous and the rules of classical mechanics can be applied. In this 'quasiclassical' regime, the electron's motion can be described both classically and quantum mechanically. The idea, now, is to prepare hydrogen atoms in quasiclassical quantum levels, apply a gentle excitation (actually either a low-frequency electric or magnetic field) of a form *known* to produce chaos in the classical domain, and to examine what happens to the quantum description. It turns out that chaos does not make itself felt in the wave function of any particular energy level, but rather in the *distribution* of energy levels. In fact, in the non-chaotic case the energy levels of the quasiclassical regime are found to be distributed randomly, while in the chaotic regime the energy distribution develops a structure. For other more complex quasiclassical systems, the primary signature of quantum chaos still appears to reside in the distribution of the allowed energy levels, but sometimes additional evidence appears to sneak into the wave functions themselves. The fingerprint of chaos is therefore undoubtedly present in quantum theory, although the details are far from clearly understood at the time of writing. Even less is currently known in the true quantum (rather than quasiclassical) context, and here the term 'quantum chaos' still, in many ways, serves more to describe an enigma than to define a well posed problem.

In spite of the problems that still remain, it is now quite clear that twentieth century physics has dealt a death blow to Laplace's determinism. Since a central dogma of quantum theory is the fact that there is a fundamental limit to the accuracy with which position and velocity can be simultaneously measured (see chapter 12), and since the studies of chaos have made it quite clear that many (one might even say most) disordered many-body systems can magnify even the most minute uncertainties into gross and wholly unpredictable variations, the conflict between determinism and free will is no more. In short, the hope that some futuristic computer, when programmed with all the data necessary to describe the world at an instant of time, could use the fundamental equations of physics to determine the future is unfounded even in principle.

Chaos is still an extremely young science and our present understanding of it undoubtedly primitive. Yet already, in only a little more than two decades, the explosion of activity in its field has crossed so many boundaries between seemingly unrelated disciplines that some are already predicting that the twentieth century will be remembered for three, and only three, truly great breakthroughs in physics—namely relativity, quantum mechanics and chaos. Already chaos possesses its own scientific journals, in which can be found contributions in the fields of engineering, physics, chemistry, oceanography, physiology, astronomy, meteorology, biology, ecology, and (no doubt) several other 'ologies' that have escaped my notice. Chaos leads to structures of infinite complexity and beauty and yet, through it all runs the essential notion that such a richness of structure in the physical world is not necessarily a consequence of the complexity of physical laws, but rather is more frequently a simple mathematical consequence of the reiterated application of extremely simple laws.

8

Symmetry: from Galois to Superstrings

Symmetry manifests itself in the physical world in many different ways, some very easy to appreciate and others quite difficult. It is somewhat unfortunate (for our normal chronological presentation of the development of the mathematics supporting important breakthroughs in physics) that the mathematical formalism that first revealed the enormous power of 'symmetry' arose in one of the less visual contexts — namely, the theory of polynomial equations. It is also true that the mathematical expression of the consequences of symmetry introduces concepts and operations that are unfamiliar to most 'amateurs'. As a consequence, it is often said that the formal theory of symmetry, or 'group theory' as it is now commonly known, is like garlic in that it is almost impossible to have a little bit of it. Although I can appreciate this sentiment, I do not think that it is necessarily true if our introduction turns away from the historical development of the subject, and our initial attention is focused upon the simplest of all possible symmetries both to envisage and to represent in mathematical terms.

The symmetry in question is a geometric one (geometric symmetries being by far the simplest of symmetries for the non-expert to appreciate) and is perhaps the most common of all symmetries in nature. It is reflection symmetry. If, for example, you hold a mirror perpendicular to this page and parallel to the long edge of the page, then most of the letters in the mirror will look strange, but a few (like M, W, o, v, and x) will remain quite recognizable. What characterizes the latter is the fact that they do not change their appearance when reflected about a line (or plane to be more precise) running down their middle. We say that they possess a reflection symmetry about this plane or, more generally, that they exhibit a 'bilateral' symmetry. So many objects around us have this symmetry in an exact (e.g. balls, cups and saucers, wheelbarrows) or approximate (cats, dogs, cars, trees, ourselves) form that when we look in a mirror it is not immediately obvious that, in the mirror, right has changed into left and *vice versa*.

In order to discuss this property of bilateral symmetry in mathematical terms it is now necessary to define the different 'operations' that transform an object possessing this symmetry into itself. There are only two of them. One is the 'stay as you are', or 'identity' operation I, and the other is the reflection operation R. You may think that the inclusion of the identity operation is a bit of a 'cheat', since it is a 'do nothing' operation but, as we shall soon see, its inclusion has important consequences in the mathematical description of the symmetry to follow.

The most important first observation about the two operations I and R is that, when they are performed consecutively in any order (that is as $I \cdot I$, $I \cdot R$, $R \cdot I$ or $R \cdot R$, where in general $X \cdot Y$ means 'first do operation Y and then follow it by operation X') they are still always equivalent to either I or R. For example,

$$I \cdot I = I \qquad I \cdot R = R \qquad R \cdot I = R \qquad R \cdot R = I$$

as you may easily verify for yourself. Since these combinations of operations do not give rise to any new operation, we say that I and R form a 'closed set'. This being the case, we can symbolically represent the above set of equations by a two-by-two 'multiplication table' as follows:

$$
\begin{array}{c|cc}
 & I & R \\
\hline
I & I & R \\
R & R & I \\
\end{array}
$$

where the interpretation is obvious from the pattern. The next important step is to notice that it is possible to replace I and R by numbers and still obey the multiplication table if we now interpret the 'dot-product' $I \cdot R$ mathematically as 'I times R' and so on. Moreover, this can be accomplished in two different ways: first by putting

$$I = 1 \qquad R = 1$$

and second by putting

$$I = 1 \qquad R = -1.$$

Each of these choices satisfies all four bilateral symmetry equations $I \cdot I = I, I \cdot R = R, R \cdot I = R,$ and $R \cdot R = I$ as required. We say that we have found two numerical 'representations' of the symmetry in question; $I = 1, R = 1$ being called the 'symmetric' representation, and $I = 1, R = -1$ the antisymmetric one.

This is all well and good, you may be saying, but what use is it? Well, for example, it tells us that any motion which takes place subject to forces that possess this I, R kind of symmetry can be of only two fundamentally different types; namely, symmetric or antisymmetric. In order to understand clearly what we mean by this, it is only necessary to choose a simple example. In figure 40(a) we show three beads on a piece of straight wire. The outside beads are of the same mass (m) while the center bead is of a possibly different mass (M), and each outside bead is connected to the center one by identical springs. It is clear from the figure that, for this system, the plane through the center bead perpendicular to the wire is a plane of reflection symmetry. This arrangement is therefore subject to the bilateral symmetry restrictions spelled out above.

Suppose that we now gently disturb the beads to set them into an oscillatory motion. The above 'group theory' tells us that there are only two possible kinds of 'simple' motion that can take place, where by 'simple' I mean periodic oscillatory motion that repeats itself endlessly (ignoring frictional slowing down). These two so-called 'normal modes' of oscillation are the symmetric and anti-symmetric modes illustrated respectively in figures 40(b) and 40(c). But how do we *know* that these are the relevant modes? In

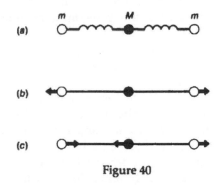

Figure 40

what sense are they symmetric and antisymmetric, and where is the connection with the group multiplication table set out above?

Look first at figure 40(b). In it, the arrows indicate the displacements of the oscillating masses at a particular instant of time. The end masses are moving equally outward and the center mass is stationary. Later on in the oscillation, the end masses will move equally inward with the center mass still at rest. This motion is the 'symmetric' mode because it has the same properties under the transformations (or operations) I and R as does the symmetric representation $I = 1$, $R = 1$. For example, when figure 40(b) is reflected through the reflection plane, it turns into itself (i.e. the operation R is equal to 1), while the identity operation I is always equal to 1 by definition. Note that in this symmetric mode the outside masses m must always be moving in exactly the 'opposite' fashion, while the center mass must always be exactly at rest. Any tiny variation from these requirements would violate the dictates of the symmetry.

Now look at figure 40(c). This is a pictorial representation of the antisymmetric mode. In this mode the outside masses move equally in the same direction, and the center mass can now also move. In fact, if we wish to keep the 'center of gravity' of the entire mode fixed, then the center mass must move in the opposite direction from that of the outside masses. And what happens if we reflect this mode in the bilateral plane of symmetry? It is easily seen from the figure that it reflects into its negative (that is, the arrows indicating the mass displacements all turn around to their opposite directions). This means, for this particular mode, that the operation R is equivalent to -1. With $I = 1$, once again, we now identify this mode as conforming with the group theoretical dictates of the antisymmetric representation; that is $I = 1$, $R = -1$.

Now, in truth, if you set up the arrangement shown in Figure 40(a), and you start the motion by just tweaking one of the end beads, then the resulting oscillations will appear to be quite different from (and much more complicated than) either of the two normal modes discussed above. However, this motion, and any other you can contrive by starting off in other ways, can always be mathematically represented as a mixture of the two normal modes in different proportions. It is only the fact that the two normal modes have, in general, different frequencies that makes these mixtures appear so complicated at first sight. It therefore follows that the symmetric and antisymmetric normal modes are the fundamental vibrations from which all actual modes can be assembled by simple combination. The essential point of this simple exercise is to make clear that the allowed vibrational symmetries of these fundamental vibrations are determined entirely by the 'multiplication table' of the symmetry group of the forces and masses involved in the motion. It is true that symmetry alone does not provide us with all we might wish to know (such as the frequencies of the normal modes, for example, which depend upon the detailed physics of the problem involving the strength of the springs and the numerical values for the masses) but amazingly the fundamental *pattern* of each normal mode motion is completely determined by 'group theory', by which we mean the manner in which the symmetry operations combine.

In the context of molecular physics, this same geometric configuration might represent a linear (that is, straight line) three-atom molecule with two different atom types m and M. Our conclusions remain unchanged, regardless of the nature of the interactions and whether they be of a simple harmonic form or horrendously complex. The important point is that symmetry considerations alone enable us to deduce certain *exact* features of what might be (in the presence of complicated forces) an extremely difficult problem to solve. This has given us the total number of independent solutions (in this case two) and has enabled us to name them (in this case by the terms 'symmetric' and 'antisymmetric') in a fashion that carries with it certain *exact* implications concerning the fundamental nature of the motion in question.

The next most simple symmetry of geometry involves (not surprisingly) three, rather than two, symmetry transformations. The geometrical object conforming with this symmetry is not, however, that which you might first guess; namely the equilateral triangle or, physically, three identical atoms at the corners of an equilateral triangle. The latter configuration, as you can quickly verify, possesses six symmetry operations: three are rotations (through 0, 120 and 240 degrees, respectively), and three are reflections (through the perpendicular bisectors of the sides). The simpler symmetry with only three independent

3-fold rotational symmetry

Figure 41

symmetry operations is, in fact, rarely found in molecular systems, but is exhibited by a rather more amusing, if less scientific,
object; namely the symbol of the Isle of Man (which is a small
island lying in the Irish Sea between Ireland and Great Britain).
This symbol is shown in figure 41, and its three symmetry
operations are the rotations through angles 0, 120 and 240
degrees, which we label I, A and B, respectively.

It is important to note that rotations of 360, 480, 600, etc are not
counted as additional symmetry operations since they are each
identical to one of I, A and B. To see this it is only necessary to
place a point at some position on the symmetry 'object' in
question which has no special symmetry significance (say, an
arbitrarily chosen point on the circular rim in figure 41). All
operations that transform this point to the same position are
identical transformations from a symmetry point of view. It is
now easy to verify the 'multiplication' properties for these three
operations as follows:

$$
\begin{array}{lll}
I \cdot I = I & I \cdot A = A & I \cdot B = B \\
A \cdot I = A & A \cdot A = B & A \cdot B = I \\
B \cdot I = B & B \cdot A = I & B \cdot B = A
\end{array}
$$

which 'translates' into the more pictorial pattern

$$
\begin{array}{c|ccc}
 & I & A & B \\
\hline
I & I & A & B \\
A & A & B & I \\
B & B & I & A
\end{array}
$$

when cast in the form of a 'multiplication table'. In order to find
the numerical representations for this pattern (that is, the associ-

ations of symbols with numbers whose multiplication in pairs gives rise to this same configuration) it is now necessary to incorporate complex numbers. There are three different combinations that satisfy the multiplication table, as follows:

$$I = 1 \qquad A = 1 \qquad B = 1$$

$$I = 1 \qquad A = a + ib \qquad B = a - ib$$

$$I = 1 \qquad A = a - ib \qquad B = a + ib$$

in which $i = \sqrt{-1}$, $a = \cos (2\pi/3) = -1/2$, and $b = \sin (2\pi/3) = \sqrt{3}/2$. Each of these representations leads to a normal mode of oscillation for any motion subject to this symmetry.

Clearly, one can now press on to geometrical objects with an increasing number of symmetry operations, like a rectangle (with four), an equilateral triangle (with six), and a square (with eight). The same expressions of symmetry can, of course, equally well be extended to three dimensions (or even, for those who can envisage such objects, to four dimensions and higher). As the number of symmetry operations (or symmetry 'elements', as they are sometimes called) increases, the problem of finding their numerical representations becomes increasingly difficult. However, for each case the mathematical task need only be performed once by one competent individual, after which the various representations can be tabulated for all to see, so that the difficulty is in practice always avoided. In general, these numerical representations require an extension of the concept of number to include not only complex, but matrix quantities (see chapter 5) and the latter point is worthy of note since it carries with it the concept of 'degeneracy'. In the context of vibrations, degenerate modes are modes of different symmetry but with identical frequencies. Sometimes, in physics, two or more modes of oscillation can have the same frequency for some 'accidental' reason (that is, a reason pertaining to the non-symmetry-related aspects of the problem). More frequently, however, degeneracy occurs for certain modes as a *requirement* of the symmetry of the problem, and is necessarily present regardless of all the other details of the system in question. Such degeneracies 'show up' in group theory as matrix representations; a three-by-three matrix representation, for example, implying the necessary degeneracy of the three modes in question. All of this can naturally find valuable application in

molecular physics and crystallography where geometric symmetries abound. But we are ahead of our story and are now in a
position to return historically to the pure mathematical problems
that first gave light of day to the notions of group theory.

In the 1760s, one of the primary interests of the French mathematician Lagrange concerned the solution of algebraic equations.
The problem at hand was a longstanding one concerning the
techniques for obtaining explicit algebraic solutions to polynomial equations like

$$ax^2 + bx + c = 0$$

$$ax^3 + bx^2 + cx + d = 0$$

$$ax^4 + bx^3 + cx^2 + dx + e = 0$$

and so on, to equations of higher degree than four. By explicit
algebraic solutions we mean obtaining a set of formulas that
express the 'unknown' x in terms of the coefficients $a, b, c, d, e, \ldots,$
etc, the formulas themselves involving only a finite number of
operations related to addition, subtraction, multiplication, division, or the extraction of roots. Indeed, the fundamental question was 'Can such a solution always be obtained, even in
principle?'

The answer for the case of the quadratic equation had been
known to be 'yes' since Babylonian times. Thus, all High School
text books will inform you that the solutions sought can, for this
case, be written

$$x = (1/2a)[-b + \sqrt{(b^2 - 4ac)}]$$

$$x = (1/2a)[-b - \sqrt{(b^2 - 4ac)}].$$

If solutions are allowed to include complex numbers, then it was
also suspected that each equation would have the same number
of solutions at the degree (that is, the largest power of x) in the
equations. For example, this assertion had been made by France's
leading mathematician during the mid-eighteenth century, Jean
Le Rond D'Alembert (1717–1783), in 1746, but only partially
proved. The full proof was given by Gauss in his doctoral
dissertation in 1799. The three solutions for the general cubic, and
the four solutions for the general quartic, had also been known
since the sixteenth century, so that the focus since that time had

been directed primarily towards the solution of the general quintic.

The common method which had worked for the equations so far mastered (an example of which was given for the cubic equation in chapter 5) concerned the transformation of the equation to one of the next-lower degree by a suitably chosen substitutional procedure. However, for the quintic equation, Lagrange found that a parallel procedure this time led to an equation of one higher degree (a sextic) rather than the hoped for quartic. He therefore conjectured that the general quintic polynomial equation just might not possess algebraic solutions at all (i.e. it would be necessary to 'solve' them graphically or numerically if at all).

In this conjecture he turned out to be correct, a fact proved in 1799 by Paolo Ruffini (1765–1822) and again, three decades later (in a more convincing fashion) by the Norwegian Niels Henrik Abel (1802–1829). Nevertheless, the basic idea that was eventually to lead on to group theory and to confirming the implied more general conclusion—that no such solution existed for any general polynomial equation of degree higher than four—was first introduced in Lagrange's battle with the quintic equation. This idea centered upon examining the solvability of equations in terms of the 'permutation' symmetry of their solutions (or 'roots' as they are called), where by permutation symmetry we mean the pattern of operations that transform one solution to another.

In order to examine this procedure in more detail let us consider a specific equation, say the cubic

$$x^3 - 3x + 1 = 0$$

the three roots of which we can call a, b, and c. Now it happens that the roots for this case are known from geometry to be

$$a = 2 \cos (4\pi/9)$$

$$b = 2 \cos (8\pi/9)$$

$$c = 2 \cos (16\pi/9)$$

and seem to suggest an angle-doubling symmetry property, since a continuation of this doubling to $2 \cos (32\pi/9)$, $2 \cos (64\pi/9)$ is easily verified to generate no new solutions but merely to regenerate the old ones. Using the double-angle formula from trigonometry (namely, cos $(2y) = 2\cos^2 y - 1$) the implied relationship between the solutions is readily seen to be

$$a = c^2 - 2$$
$$b = a^2 - 2$$
$$c = b^2 - 2.$$

Now, although we obtained these relationships via the known solutions, it is quite possible to generate them without knowing the solutions at all and such methods are available for equations of any degree. Although these methods are too difficult to dwell on here, it is easy to verify (by direct substitution; see appendix 3) that the above relationships are valid for the cubic equation in question without reverting to the actual solutions. This, at least, gives credence to the idea that symmetry relations can be verified without a direct knowledge of actual solutions. For our particular equation, these three relationships embody within them, in some fashion, a symmetry of the equation itself.

In order to fully appreciate this symmetry, we now ask what 'procedure' generates the second relationship from the first, the third from the second, and so on? The answer is that in the first case we replace c by a, and a by b. Correspondingly, in the second case we replace b by c, and a by b; and in the third case (if we wish to regenerate the first relationship from the third) we replace c by a, and b by c. These three transformations, namely $a \rightarrow b$, $b \rightarrow c$, $c \rightarrow a$, are referred to as 'cyclic permutations' and are often expressed by the 'shorthand' notation

$$w = (abc).$$

The 'operator' w acting upon any function $f(a, b, c)$ of the parameters a, b, and c therefore transforms it to $f(b, c, a)$. This operation can, of course, be repeated (w^2) to generate $f(c, a, b)$, or applied three times (w^3) to regenerate the original $f(a, b, c)$. If we label $w = A$, $w^2 = B$, and $w^3 = I$, then we find that these operations form exactly the same multiplication table as did the geometric three-fold symmetry operations on the Isle of Man symbol of figure 41. Technically known as the 'cyclic group of order three' it follows that this same group is the symmetry group of the equation $x^3 - 3x + 1 = 0$, in which context it is known as the 'Galois group' of this equation, after the youthful Evariste Galois

Evariste Galois, 1811–1832. Picture from *Magasin Pittoresque* by his brother Alfred. Reproduced by permission of Mary Evans Picture Library.

(1811–1832) whom we shall meet again later. The Galois group therefore consists of those interchanges that can be made between the roots of an equation such that any polynomial statement relating them which is true before the interchanges are made is still true after the interchange.

This idea, that all equations can be associated with patterns of symmetry operations, lay relatively dormant after Lagrange's battle with the quintic; dormant, that is, until 1830, when the 19-year-old French mathematical prodigy Evariste Galois took up the quest. Young geniuses whose lives are cut short by death from duelling are part of the literary tradition of the Romantic Age. But in Galois we have a real-life example. When he first entered school at the age of 12 he showed little interest in Latin or Greek but was fascinated by mathematics, particularly geometry,

although his routine classwork remained mediocre and his
teachers regarded him as merely eccentric. By the age of 16 Galois
knew what his teachers had failed to recognize—that he had a
truly exceptional, if unconventional, mathematical mind.
Eschewing routine preparation, he failed time and again the
entrance examinations to distinguished schools. His papers sub-
mitted to the French Academy in prize competitions were also
passed over and, thoroughly disillusioned, he eventually joined
the National Guard. After one or two arrests for political offenses,
he eventually became involved in a 'duel of honor' over a young
lady. The night before the duel, fearing that he might not survive,
Galois spent the time jotting down notes for posterity concerning
his mathematical discoveries. He asked that the letter be published (as it was within the year) and expressed the hope that the
premier mathematicians of the age might give some consideration to the importance of his theorems. On the morning of 30
May 1832, Galois met his adversary, and died the following day as
the result of his wounds. He was 20 years old.

Galois' achievement, in the context of the solution of polynomial equations, was not only his demonstration of the fact that
any such equation could be associated with a 'symmetry group',
but that these symmetry groups could be separated into two
kinds called 'reducible' and 'irreducible'. Reducible groups are
those which contain operational 'multiplication' patterns that can
be separated (in a sense too involved for us to pursue here) into
the 'product' of two or more smaller groups. When an equation
can be broken up into simpler equations, this fact always betrays
itself in the pattern of the group multiplication table, the pattern
splitting up into recognizable 'blocks'. One such example is the
table for the symmetry group of the equilateral triangle (which is
also the Galois group of the equation $x^3 = 2$). It is made up of a
six-by-six pattern that can be cast in the form

$$X \quad Y$$
$$Y \quad X$$

where both X and Y are three-by-three matrices. Galois showed
that the general polynomial equations of fifth or higher order
possessed symmetry groups that could not be reduced to the
degree necessary to permit an algebraic solution. Note, however,

this does not imply that there are not *some* equations of higher degree which can be solved in this fashion.

It took several decades after Galois had set out his original ideas for other mathematicians to fully appreciate them. Although his untimely death contributed somewhat to this situation, he was not the most articulate of mathematicians and, even after his works were published posthumously, they proved to be difficult for his contemporaries to digest in detail. And if it took the mathematical world a few decades to grasp the importance of group theory, it is perhaps not surprising that it took the scientific community a full century before the immense utility of these concepts, born in efforts to solve polynomial equations, were grasped. Nevertheless, by the 1840s, the basic notion was solidified that certain 'operations' could be defined (permutations in Galois theory) that would transform a set of arithmetic 'elements' (solutions of equations in Galois theory) in a manner that would be completely 'closed' in the sense of creating no new elements outside the set. Moreover, this could be done in a manner more general than that associated merely with the solutions of algebraic equations. As the mathematical study of the various aspects of symmetry progressed, it became clear that the concept of a 'group' of operations (that captured the mathematical 'soul' of this symmetry) could be defined in a completely abstract manner. There are only four rules:

1. A 'group' is a set of 'elements' $A, B, C, \ldots$, etc which can be combined with one another to obtain other elements. The combination of any two elements $A \cdot B$ is also a member of the group.

2. One of the elements I is a 'stay as you are' or 'identity' element, so that $A \cdot I = I \cdot A = A$ for all elements A in the group.

3. For each element A in the group there is also an 'inverse' element A^{-1} such that $A \cdot A^{-1} = A^{-1} \cdot A = I$.

4. For all elements $A, B, C, \ldots$ in the group $A \cdot (B \cdot C) = (A \cdot B) \cdot C$, where the brackets mean 'do that combination first'.

The furthering of such ideas was espoused particularly by Cauchy in his probing of the possible permutations and rearrangements of a finite number of different objects. In a more analytic vein this led, by the 1880s, to the concept of a number field with a finite number of elements. The formulation of a finite number field, though perhaps rather daunting when presented in its abstract generality, can be made almost trivially simple by means of examples. Consider, for example, the process of 'counting' as performed by a clock. According to the hour hand, three (hours) added to eleven (o'clock) makes two (o'clock) or, in symbols, $3 + 11 = 2$. Similar 'clock equations' like $4 + 12 = 4$, $3 + 7 = 10$, and $1 - 2 = 11$ are easily created. More generally, this clock arithmetic (or modular arithmetic, to give it its proper name) can be carried out not only on a '12-clock', but on clocks that use a counting base unit different from 12. A general 'n-clock' then records only remainders after all sets of n have been 'forgotten'.

Let us now consider in detail an extremely simple example—say a '3-clock'. The '3-clock' only has three integers on its 'face', namely 0, 1 and 2 (3 and 0 being the same 'number' on such a clock). Counting 'modulo three', to use the proper terminology, we quickly verify that

$$0 + 0 = 0 \qquad 0 + 1 = 1 \qquad 0 + 2 = 2$$
$$1 + 0 = 1 \qquad 1 + 1 = 2 \qquad 1 + 2 = 0$$
$$2 + 0 = 2 \qquad 2 + 1 = 0 \qquad 2 + 2 = 1$$

covers the entire range of possible additions. Since $4 = 1, 5 = 2, 6 = 0$, etc when counting (mod 3), larger integers never appear. It follows that within the 'operation' of addition, we can transform the 'elements' 0, 1, 2 into each other in a manner that introduces no new elements. The pattern of numbers so created can be displayed in the tabular form

+	0	1	2
0	0	1	2
1	1	2	0
2	2	0	1

where comparison with the above equations makes the interpretation self-evident. The plus sign in the top left-hand corner reminds us that the operation combining the numbers for this table is addition. This is important because we can now also construct an analogous pattern for which the combining operation is multiplication.

Multiplication on a '3-clock' looks like

$0 \times 0 = 0$	$0 \times 1 = 0$	$0 \times 2 = 0$
$1 \times 0 = 0$	$1 \times 1 = 1$	$1 \times 2 = 2$
$2 \times 0 = 0$	$2 \times 1 = 2$	$2 \times 2 = 1$

which translates to the tabular form

×	0 1 2
0	0 0 0
1	0 1 2
2	0 2 1

We may now verify that we have produced a system with only three 'elements'—namely, 0, 1, 2—in which all the usual rules of · arithmetic (involving addition, subtraction, multiplication and division) hold. In particular, division follows from the multiplication table by the normal rules of arithmetic. Thus, 1 divided by 2 (or $\frac{1}{2}$) is the number which, when multiplied by 2, gives 1. The × table tells us that this number is 2. In modular arithmetic this is written as

$$\tfrac{1}{2} = 2 \ (\text{mod } 3).$$

Mathematicians call sets of numbers that possess all these properties 'fields'. In other words a number field is a set of numbers that is closed under all the four arithmetic operations $(+, -, \times, \div)$. The most familiar number fields for everyday use are infinite—such as the set of all fractions. But we now have examples of finite number fields. Note, however, that only 'n-clocks' with n equal to a prime number give rise to number fields. What goes wrong in other cases? Try a '4-clock' system and check it out for yourself.

With these definitions we now see that the elements 0, 1, 2 of the '3-clock' do form a group under the operation of addition (that is, $A \cdot B = A + B$) but not under multiplication ($A \cdot B = A \times B$). Can you spot which group law (or laws) the multiplication table violates? We also note that the pattern of group symmetry elements for the '3-clock' addition table is that same old pattern that appeared earlier for the geometric symmetry of figure 40 and for the Galois symmetry of the equation $x^3 - 3x + 1 = 0$. A suspicion is beginning to arise that there is, perhaps, only one pattern of symbols that exists for a group of three elements regardless of their meaning, and such is indeed the case. And what about other groups with a finite number of elements; are they unique? For two elements the answer is yes (the bilateral group), but for groups of order four (i.e. with four elements) the answer is no. There are two different groups of order four, and the geometric symmetry of the rectangle and the addition symmetry of the '4-clock' provide examples of them. Check it out for yourself. Simple finite groups therefore seem to have an 'existence' over and above any specific examples which might exhibit their particular symmetry and this is one of their charms. The number of different groups of order n begins as follows:

Order	Number of groups	Order	Number of groups
1	1	7	1
2	1	8	5
3	1	9	2
4	2	10	2
5	1	11	1
6	2	12	5

Each group is a unique mathematical object. However, their ultimate significance goes even deeper than this because, just as in arithmetic every integer possesses a unique factorization into primes (e.g. $60 = 2 \times 2 \times 3 \times 5$), so every group can be 'factored' or 'reduced' in a certain sense (related to that discussed before in connection with the group of operations for the equilateral triangle). The 'prime numbers' of group theory are therefore those groups that can be factored only into a product of themselves with the single identity element. We refer to them as 'simple' or 'irreducible' groups and a complete classification of them for all finite groups has only just been completed (in the 1980s).

The concept of groups, originated and named by Galois and honed by Cauchy, inspired mathematicians to create completely new algebras during the second half of the nineteenth century. These notions were first gathered together into a full-blown branch of algebra (what we should nowadays refer to as 'modern algebra') by the German mathematician Felix Klein (1849–1925)

Sophus Lie, 1842–1899. Photograph taken in Göttingen ca 1870. Courtesy of Elin Strøm Historisk Institutt, University of Oslo, and Institute of Mathematics, University of Oslo.

who spent much of his life developing, popularizing, and applying group theory. In fact, so contagious was his enthusiasm for the subject that many, in the late-nineteenth century, prophesied that all of mathematics would one day be comprised within the theory of groups. This attitude was encouraged by the demonstration in the late 1880s that groups need not be restricted to symmetries with a finite number of elements. The extension to groups of infinite order was first accomplished in the study of so-called 'continuous transformations', and was primarily the work of an old fellow student of Klein's, the Norwegian Sophus Lie (1842–1899). Lie studied the symmetry exhibited by objects such as a circle or a sphere. Evidently each is unchanged by any rotation whatsoever of the coordinate axes defining them—even an infinitesimal rotation.

Think, for a moment, about the mathematical description of a coordinate transformation in elementary geometry. For simplicity we may focus on rotations in a plane. A counterclockwise rotation of the (x, y) coordinate frame about the origin $(0, 0)$ through an arbitrary angle θ is a symmetry operation for any system exhibiting circular (or axial) symmetry. Using elementary trigonometry, we see that this operation transforms the coordinates of a point P from (x, y) to

$$(x \cos \theta + y \sin \theta, \ y \cos \theta - x \sin \theta).$$

The complete set of these rotations forms a group of infinite order since θ is a continuous variable. If we first perform a rotation through an angle θ_1, then follow it with a second rotation through an angle θ_2, the result is just a rotation through $\theta_1 + \theta_2$. This gives the rule for combining rotations and it is not difficult to show that all the required properties for defining a group are satisfied. Continuous groups of this kind are now known as 'Lie groups' in honor of their discoverer. The above group for two-dimensional rotations is usually denoted as R_2, and it is fairly self-evident that it must have analogous counterparts in higher dimensions (R_3, R_4, ..., etc).

Thus, by the end of the nineteenth century, the complete mathematical edifice of group theory was essentially established. But all of this was far removed from the realm of most of the physicists and chemists of the day. In science, group theory was largely ignored with one important exception—namely, the field of crystallography, and the work (set out in chapter 2) leading to the complete determination of all the allowed symmetry structures for crystals. This work became particularly important in the 1920s, at which time the study of atom positions in crystals, using the diffraction of x-rays, finally got underway in earnest. It was not so much that crystallographic group theory was useful in labeling the structures once they were determined, the essential impact of the theory came in interpreting the experimental diffraction patterns in order to extract the structure. When an x-ray is scattered by an atom, it can leave at any angle with respect to its incoming direction. However, if the atoms are arranged in some ordered configuration, the scattered x-ray waves either tend to reinforce or interfere with one another depending on their directions relative to the spatial symmetries of the atomic arrangement in the crystal. Using group theory, it is then possible to calculate which spatial directions are allowed or disallowed for any particular crystal structure and these results are *exact* even if the precise mechanism by which the atoms interact with the x-rays is unknown. In this manner the observed diffraction patterns betray the crystal structures under study.

The first significant use of infinite order (or Lie) groups in physics came with the spectroscopic study of electrons in atoms. These spectral patterns, produced when light or x-rays are absorbed by atomic electrons, were enormously complex and for years defied interpretation. The breakthrough came with quantum theory together with the realization that an electron 'orbiting' about an atomic nucleus experienced a force field of spherical symmetry (R_3). This is exactly true for the hydrogen atom (which contains only a single electron) but is also true to a good approximation for many-electron atoms since, for them, the symmetry-breaking interaction between electrons is essentially 'smoothed out' by their relative motion.

In quantum theory an electron in an atom cannot be pictured as an orbiting particle but as an electron cloud, the symmetry of which, we now suspect, is largely controlled by the Lie group R_3. Quantum theory also restricts the electron energy to an infinite sequence of discrete values. With each one is uniquely associated a wave function that determines the probability distribution of the electron in space for this level. Naturally, the simplest numerical representation for R_3 is the one that associates the number 1 with all rotation elements. This representation is non-degenerate and is present for all symmetry groups, finite or continuous. For the atomic problem it corresponds to the lowest (most-stable) energy, is called the s state, and is associated with a wave function (or electron cloud) of full spherical symmetry (see figure 42(a)). The next-highest energy level can be shown to require 3×3 matrices for its representation. It is therefore three-fold degenerate. Each of the three associated electron clouds has dumb-bell-shaped lobes along a Cartesian axis. They are known as the p states, p_x, p_y, and p_z. Sketches of these wave functions (also known as 'atomic orbitals' in the present context) are shown in figure 42(b). From the figure we note that the two lobes of each p wave function have opposite relative signs (the associated electron density, which is given essentially by the square of the

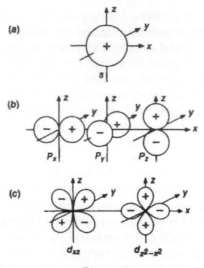

Figure 42

wave function, is naturally always positive). This sign dependence is important since it establishes that the p functions are
'antisymmetric' under 'inversion' symmetry $(x, y, z) \rightarrow (-x, -y, -z)$. More important still is the fact that *all* the allowed wave
functions are either 'symmetric' or 'antisymmetric' under this
symmetry group (the s states are obviously symmetric). This
inversion group is a two-element 'factor-group' of R_3; it must
therefore possess the same symmetry table as the reflection
group with which we introduced the group concept. In the
present context it is said to define the 'parity' of the wave
function; even-parity corresponding to the symmetric representation and odd-parity to the antisymmetric one. For higher
energy levels the group theory of R_3 continues to predict the
form, degeneracy, and parity of the orbitals involved. After the p
orbitals come the five-fold-degenerate even-parity 'd levels' (two
of which are shown in figure 42(c)), and so on, and so on.

The symmetry properties of these atomic orbitals (and particularly their parity) can now be combined with the symmetry
properties of the electric and magnetic field components of light
or x-rays to determine which pairs of atomic levels can be
'excited' by the relevant radiation and which cannot. It is these

excitations (representing transfers of electrons between the energy levels concerned) that physicists observe when probing the atom spectroscopically, and it is therefore quite evident that group theory has an enormously important role to play in unraveling the spectral details. And although electron–electron interactions within the atom slightly perturb some of the predicted degeneracies in atoms more complex than hydrogen, the essentials of the group theoretical description remain. Although it remains true, as was the case for the bilateral symmetry problem of figure 40, that symmetry (or group theory) cannot by itself solve the entire problem as regards all numerical details, by far the most important attributes of atomic orbitals follow from an understanding of R_3. In particular, the symmetry-determined directions of the orbital lobes of the various allowed wave functions prove to be of immense value in determining the nature of directional bonding in chemistry.

Progressing from atoms to molecules, the basic story can be repeated. Now, however, it is necessary to work using the symmetry group of the molecule in question rather than R_3. The simplest case, that of a molecule containing only two atoms, is invariant under the Lie symmetry group for axial symmetry, namely R_2. For this group all the representations are one-dimensional so that no degeneracy exists. More complicated molecules, such as those forming triangular configurations, are invariant only under the symmetry operations of finite groups; but here too, group theory proves to be invaluable in deducing the symmetry of the 'allowed' electron orbitals (now called molecular orbitals) and for analysing their interactions with electromagnetic radiation. Finally, progressing further from molecules to crystals, we get back to the point groups and space groups discussed in chapter 2. In this context group theory enables us to examine the manner in which the discrete electronic energy levels of the individual atoms are perturbed by the interactions between atoms as the crystal is considered to be assembled or 'brought together' from an initially widely separated collection of atoms to the actual equilibrium configuration and atomic spacing of the real stable crystal. The sharp individual atomic energy levels (which result from solving the isolated atom problem) are found to broaden out into bands. The translational invariance (also a group symmetry property) allows the reintroduction of the 'wave vector' concept, first seen by us in chapter 4, in terms of which many of the details of these 'band structures' can be detailed. The latter, in particular, decide whether the crystal in question is to be a metal, semiconductor or electronic insulator.

However, in modern physics, the utility of group theory extends far beyond the electronic properties of atoms (in or out of solids) even to the far far smaller domain of the atomic nucleus itself. An arbitrary nucleus can be viewed as a stable state of many interacting protons and neutrons (collectively referred to as nucleons). These nuclear 'particles' inside the atomic nucleus are not packed tightly together but actually move about quite freely in almost independent orbitals. However, the differences between electron orbital motion and proton/neutron orbital motion are profound. Most significantly there is no massive central object at the center of the nucleus around which the protons and neutrons orbit. Nevertheless, each particle within the nucleus can be considered (with reasonable accuracy) to move in the average force field of its neighbors and, within this approximation, they are then each subject to the rotational invariance of the Lie group R_3. Although the actual forces involved inside the nucleus (the so-called nuclear 'strong' forces) are of a quite different physical origin from their electronic counterparts and require a completely different mathematical description, the same group theoretical rules of R_3 symmetry apply. Once more the allowed shapes of the probability distribution clouds for the particles (this time nucleons) can be determined. The result is a quasi-atomic model for the nucleus itself, called the nuclear shell model. Using this picture makes it possible to interpret a tremendous wealth of experimental nuclear phenomena. Collective modes of oscillation of the nucleons can be derived in terms of which it is possible to discuss the dynamics of nuclear deformation, rotation, and vibration, together with all sorts of combinations of these gymnastics. Again, group theory cannot furnish us with all the information we might wish to have, but it does enable us to break up the problem in a manner that reduces the complete motion to a number of simpler ones. Each of these motions can be studied separately subject to the symmetry restrictions present.

All the symmetries discussed so far in association with physics have been of a geometrical kind. However, in nuclear physics, and even more so in elementary particle physics (involving the study of subatomic particles like quarks), other kinds of symmetries are thought to play an essential role. For example, as the renowned German theoretical physicist Werner Heisenberg first recognized in 1932, protons and neutrons can be considered to be two different states of the same entity, the nucleon. Thus, although they differ in electromagnetic properties (the proton having a positive charge while the neutron is electrically neutral with zero charge) they are completely indistinguishable in terms of the nuclear (or strong) force which binds them together in the nucleus. Hence, Heisenberg attributed to the nucleon a new *internal* degree of freedom, later called its 'isospin'. The word 'spin' was included because the symmetry group involved, called SU(2), was the same as that obeyed in real space by the quantized rotations (or 'spins') of electrons. This symmetry is rather an odd one. If an ordinary spinning body is rotated in space through 360 degrees it returns to its original configuration. An intrinsic particle spin, however, does not do this. If a particle like the electron is rotated through 360 degrees it assumes a quantum state which, in its interactions with another particle, has measurably different properties from those of the starting configuration. To return a quantum spin to its original configuration requires a 'double rotation' through 720 degrees. Clearly, properties like spin somehow see a bigger world than we do, one that is very difficult for us to picture, but one which experiment tells us must be there nonetheless.

The concept of isospin is extremely valuable since it turns out that nucleons (that is protons and neutrons) are not the only particles subject to the symmetry restrictions imposed by the strong nuclear interactions. Particles subject to this strong force are called 'hadrons' and now number over 400—a very significant increase in complexity since the very first non-nucleon hadron (the so-called pi-meson) was discovered in 1947. A classification of this now very large set of hadrons has proved to be possible only with the help of group theory via the concept of isospin and another similar discrete internal symmetry property known as hypercharge. After a number of unsuccessful attempts, Murray Gell-Mann, of the California Institute of Technology, discovered in 1962 how to merge the two symmetries into a new internal symmetry group (called SU(3)) in a manner that enabled a complete classification of the then-known hadrons to be obtained. Known as the 'eight-fold' way, the theory led to a number of predictions, such as masses, magnetic properties, and decay mechanisms for the particles involved, and to the prediction of a new particle, the 'omega', whose discovery was triumphantly pronounced in 1964. The theory, however, was still not complete, and various extensions were subsequently examined, with hadrons being thought of as bound states of even smaller particles known as quarks (with three quarks per hadron). Although a full understanding of the 'quarkian' nature of hadrons has yet to be achieved, it is clear that symmetry concepts (in the guise of group theory) continue to play an essential role. And this is a little surprising since *a priori* one would perhaps expect little relation between the mathematics of group theory and the physics of fundamental forces and elementary particles. However, in reality, there seems to be a surprising connection; one might almost say that nature appears to be organized using symmetry groups.

Although in daily life nature appears to display a whole range of different kinds of forces, they can, in fact, all be reduced to just four basic types. Two of these have been well understood at the classical level for a long time; gravity since the time of Newton, and electromagnetism since the time of Maxwell. Gravity is a universal force acting between all particles by virtue of their mass. Electromagnetic forces act only between electrically charged particles but, like gravity, they are of long range (obeying an inverse-square distance law) and are therefore readily apparent to our senses. The two other forces are of subnuclear dimension and have therefore revealed their presence to us much more recently via atom-smashing experimentation. They are the nuclear-binding strong force, referred to above, and the so-called 'weak' force which is responsible for radioactivity (and manifests itself more by transmuting particles rather than by pushing or pulling them).

One of the most promising recent developments of the use of group theory in particle physics has arisen in an effort to unify the four fundamental forces of nature. It involves a non-geometric 'gauge' theory, the nature of which is best introduced by considering it in a more mundane context. Consider, for example, the work necessary to lift a massive object. This work is measured as the weight of the object in question times the *difference* in height between the start and endpoint of the 'lift'. It is quite inconsequential whether we measure height from laboratory floor level, sea level, or any other level, the work depending only upon the height difference, which is independent of the choice of 'zero height'. In addition to this, the actual path taken in raising the mass from its initial position to its end position is equally inconsequential. An exactly similar situation exists for electromagnetic (rather than mechanical) work, with voltage now playing the role analogous to height. Obviously the invariances with respect to path detail and choice of an absolute zero define, for each case, a symmetry, called 'gauge symmetry'. It is a rather more abstract symmetry than a geometric one since our senses cannot immediately perceive it. Nevertheless it is undoubtedly present, and gauge symmetries are now playing a central role in the search for a quantum theory that will unify the four forces of nature.

When applied to electromagnetic forces, gauge theory leads to a consistent theory of electrons and their positively charged counterparts called positrons. The theory is called quantum electrodynamics (or QED for short) and it has been extremely successful. In particular, applied to electrons it gives the whole of classical electromagnetic theory. Central to this theory is the notion that electrons and positrons interact by exchanging photons, which are the quantum manifestations of light. This success was followed by a gauge theory for the weak force, and by its amalgamation with QED into a combined 'electro-weak' theory. The predicted 'messenger particles' for the weak force (which play a role analogous to that of the photon in QED theory) are called W and Z particles and were first detected experimentally in 1983. Inspired by this success, nuclear theorists pressed on in their attempts to integrate the strong forces equally into the picture, with messenger particles called 'gluons' carrying the strong force between quarks. Begun in the 1970s and refined throughout the 1980s the resulting combined electro-weak–strong theory became known as the 'grand unified theory'. The grand unified theory remains, at this writing, far less well established than its electro-weak predecessor since many of its fundamental predictions, like the existence of a massive particle (called the magnetic monopole) carrying the fundamental quantum unit of magnetic 'charge', and the instability of the proton (with a mean predicted lifetime of some 10^{32} years) have yet to be confirmed. Until at least some of these predictions are verified, the future of grand unified theories remains somewhat open but, in spite of this, the role played by gauge symmetries in understanding the basic forces of nature now appears to be quite firmly established.

With promising theories of three of the four forces of nature in relatively good shape, the conspicuous odd-man-out is gravitation. Although the first of nature's forces to receive a quantitative mathematical description at the macroscopic level (by Newton) it continues to resist attempts to provide it with a quantum description, in spite of the presence of a gauge-invariance. A straightforward extension along the lines of the development of quantum electrodynamics fails with the appearance of many infinite terms in the equations. So long as gravity remains unquantized (even though the effects of such a quantization have essentially no consequences at the length scale of particle physics) an embarrassing inconsistency remains at the heart of physics.

Today, most physicists are pinning their hopes on a new theory that hopefully will enable gravity to finally be united with the other three forces of nature in a quantized super-unified theory. These 'theories of everything', as they are commonly called, are currently occupying the attention of a small army of theorists. More formally they are known as 'superstring' theories; 'super' because they incorporate a supersymmetry as discussed below, and 'string' because the mathematical difficulties that seem to be ever-present when elementary particles like electrons are taken to be point-like are removed when they are thought of as tiny strings with lengths of order 10^{-32} cm (to be compared with a nuclear size some 10^{19} times as large).

It has long been known that the precise value of a particle's intrinsic angular momentum (or spin) also plays a crucial part in determining its physical properties. Particles with spin $S = \frac{1}{2}, \frac{3}{2}, \frac{5}{2},$ etc (in quantized units of angular momentum) are called 'fermions' and, as mentioned above, possess an unusual rotational symmetry in that they return to their original state only after a double revolution of 720 degrees. Particles with spin $S = 0, 1, 2,$

etc, on the other hand, are called 'bosons' and they possess the more familiar rotational symmetry which returns to an original state after a single revolution of 360 degrees. As discussed in chapter 5 in connection with quaternions, ordinary rotations in three dimensions, unlike their two-dimensional counterparts, do not commute. That is to say, the order in which one performs two rotations, say R_1 and R_2, *does* matter such that $R_1 R_2 - R_2 R_1$ is not zero. The combination of rotations $R_1 R_2 - R_2 R_1$ is called the 'commutator' of these two rotations and, using it, it is possible to construct an algebra of rotations that is valid for all 'ordinary' objects like books and bosons. The commutator relationship that is valid for the peculiar geometric properties of 'double-rotation' fermions is found to involve the 'anticommutator' $R_1 R_2 + R_2 R_1$. The algebra of fermions is therefore based on a different geometry from those of bosons. The new supersymmetry attempts to unite the symmetry of bosons and fermions in a common mathematical framework which, however, requires an increase of the number of spacetime dimensions from the four (x, y, z, t) familiar to our normal senses. The purpose of the extra dimensions is to accommodate the peculiar geometrical properties of fermions, so that these extra dimensions are not space or time in the normal sense.

Supersymmetry, which associates fermions and bosons as two 'rotational' states of the same superparticle is (albeit in a somewhat unusual form) a geometrical symmetry. Since Einstein's general theory of relativity is also geometrical in form, it was discovered that supersymmetric geometry could also be used as the basis of a supergravity theory. In this manner, it was now hoped, could the four forces of nature finally be unified, each being the manifestation of one aspect of a single supersymmetric superforce. But the final details of the theory have yet to be agreed upon and, in particular, the physical nature of the ultimate constituents of nature, the elementary particles, is still argued. In vogue at the moment is the supersymmetric form of string theory. In string theory the elementary particles are pictured as modes of vibration of the tiny strings, and the extra dimensions of supersymmetry as being 'curled up' in some fashion about the strings in loops so tiny that our senses cannot detect them. The electron, the photon, the various gluons and all the countless other known particles are, in this picture, just different harmonic vibrations of one string. The theory is still difficult to visualize, containing as it does so many dimensions more than the four that are apparent to most of us and kinds of symmetries that would have confounded even Galois himself, but theorists speak eloquently of its incredible beauty and it just may be the 'theory of everything'. Whether or not this is so, it seems now almost certain that the final quantitative explanation for all the forces and particles of nature will, whenever it is confirmed, be found to rest squarely upon the foundation pillars of group theory.

9

From Coin Tossing to Entropy

Games of chance have probably been around since prehistoric times. For example, it is not difficult to picture some less-than-heroic specimen of early man attempting to avoid 'drawing the short straw' in a desperate attempt to avoid some unpopular duty. According to Greek tradition, dice were invented to relieve the boredom of Greek soldiers waiting to invade Troy (which would be in the early twelfth century BC according to modern day estimates of the most probable date of the Trojan War) while there is irrefutable evidence that they were certainly in fashion in First Dynasty Egypt (circa 3000 BC). Round coins were also in existence in the sixth century BC in the Orient, and it is hard to imagine that they were never tossed to decide some issue involving a choice. With activities of this kind, questions concerning probability are more than likely to have arisen in antiquity. It therefore comes as some surprise to learn that the earliest writings on probability theory (these being devoted primarily to dice) date only from the 1500s AD.

Questions about chance (or probability if you like) are fraught with danger—danger, that is, of getting the answer wrong. Perhaps the simplest of all non-trivial questions concerning coin-tossing already highlights the kind of difficulties to be encountered and also gives us a little historical perspective. As late as the mid-eighteenth century it was still giving trouble to Jean le Rond D'Alembert (the eighteenth century mathematician, philosopher and encyclopedist whose work marked an epoch in the science of mechanics) in the form 'What is the probability of obtaining at least one head in two tosses of a fair coin?' Since a 'fair' coin is, by definition, one that falls heads or tails with equal likelihood when properly tossed, D'Alembert reasoned that, with only three possible outcomes (namely two heads, two tails, and one head and one tail) two of which meet the requirement, the answer should be two-thirds. So confident was he of this that he even included the statement in his encyclopedia of 1754. But was it correct? Probably not, you may well be reasoning, or the story would hardly be worth telling—so let us continue to find out why.

The best way to develop a healthy skepticism concerning D'Alembert's pronouncement on the subject is to toss a few coins for yourself. If you do, you quickly notice that the 'one head one tail' combination persistently occurs more frequently than either two heads or two tails. Why D'Alembert himself apparently never made the effort to experiment only indicates the low esteem in which practical experimentation was held compared with the 'purity' of intellectual reasoning and mathematical argument in D'Alembert's time and earlier (a point we have stressed in previous chapters). But why is the head–tail combination preferred over the others—and by how much? Well, suppose that I toss a pair of coins onto a table and that one falls to the ground to be temporarily out of my sight. I look at the coin remaining on the table and see that it is, say, a head. What do I now know? Since there is an even chance that the coin on the floor is a head or a tail, I know that I have, with equal probability, either two heads or a head and a tail. The tail–tail combination can already be excluded. And what if the coin on the table is a tail? Then, obviously, I can immediately rule out two heads, leaving two tails and a head and a tail as my two equally probable outcomes. Thus, no matter what I observe on the table, the head–tail combination always remains a possibility, while one of the other two choices is excluded. The head–tail finding is obviously favored by a factor of two over either of the other possibilities.

Staying with the table–floor 'model' of the problem, it is clear that there are four equally probable outcomes. We may label them, writing the table coin first and the floor coin second as

(head head), (head tail), (tail head), (tail tail),

or, in symbols, with H = head and T = tails, as HH, HT, TH, and TT. What this tells us, in a more general context, is that the *order* of the symbols H and T is important. That is, when two coins are tossed sequentially (or the same coin twice if you prefer), the 'head first, tail second' possibility is just as likely to arise as is the 'tail first, head second' alternative, and each will occur with the same probability as either of the other two possible results, namely 'head head' and 'tail tail'. Of these four equally probable outcomes, three satisfy the requirement of 'at least one head'. The answer that D'Alembert should have given is consequently $\frac{3}{4}$ and not his encyclopedic $\frac{2}{3}$. But let us not be too critical of D'Alembert. He was only one of several esteemed mathematicians who floundered on problems of this kind in the eighteenth century and I have little doubt that 'howlers' of similar ilk are still being made in mathematics classes (and even research papers) to this day.

Looking at the arrangement of head–tail combinations that arise in the coin-tossing exercise, we note an obvious association with the right-hand side of the simple algebraic product (or multiplication)

$$(H + T)(H + T) = HH + HT + TH + TT$$

or, in a more common notation,

$$(H + T)^2 = H^2 + 2HT + T^2.$$

Interpreting H^2 as 'two heads', HT as 'a head and a tail', and T^2 as 'two tails', we can now include the notion of probability by rewriting the equation one more time (dividing each side by 4) as

$$(\tfrac{1}{2}H + \tfrac{1}{2}T)^2 = \tfrac{1}{4}H^2 + \tfrac{1}{2}HT + \tfrac{1}{4}T^2.$$

This we read as 'if a head and a tail' can occur with equal probability of $\frac{1}{2}$ in a single coin toss (i.e. inside the bracket) then the probability in two tosses (the exponent of the bracket) of getting two heads (H^2) or two tails (T^2) is $\frac{1}{4}$, while the probability of getting one head and one tail (HT) is $\frac{1}{2}$. The total probability, of course, adds up to one, which implies certainty (in the sense that one of the combinations mentioned must occur).

We appear to have stumbled upon an algebraic method for giving us the correct answer for the tossing of two coins. The obvious question now is 'Does the method still work for more than two tosses?' Let us check it out for the next simplest case of three tosses. Paying attention to the order of the tosses, there are eight (or 2^3) equally possible outcomes as follows:

$$HHH, HHT, HTH, HTT, THH, THT, TTH, TTT$$

made up of one case of three heads (H^3), three cases of 'two heads and a tail' (H^2T), three cases of 'one head and two tails' (HT^2), and one case of three tails (T^3). In terms of probabilities this implies one chance in eight of getting three heads or (independently) three tails, and three chances in eight of getting two heads and a tail or (independently) one head and two tails. And what does the

algebraic formalism predict? Well, since by straightforward algebraic multiplication

$$(H + T)^3 = H^3 + 3H^2T + 3HT^2 + T^3$$

we find immediately, by dividing both sides by eight, a probability prediction

$$(\tfrac{1}{2}H + \tfrac{1}{2}T)^3 = \tfrac{1}{8}H^3 + \tfrac{3}{8}H^2T + \tfrac{3}{8}HT^2 + \tfrac{1}{8}T^3$$

which is correct once again.

Why is this happening and can the result be generalized
further to include any number of tosses? Well, the pattern of
coefficients that are generated by the expansion of a 'binomial'
like $(H + T)^n$, with $n = 0, 1, 2, 3, 4, \ldots$, (where *binomial* refers to
the presence of just two variables inside the bracket) is certainly
not difficult to continue to higher powers n. Although known for
many years before his time, it is commonly referred to as 'Pascal's
triangle', after the French philosopher, physicist and mathematician Blaise Pascal (1623–1662), and begins

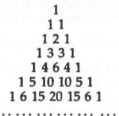

Although not originated by Pascal, it was he who, in an exchange
of letters with the finest number theorist of his day Pierre de
Fermat, first stressed the triangle's importance in the laying
down of probability theory. In association with problems involving chance, Pascal made extensive use of this triangle of numbers
to the extent that his and its names became forever associated.
The manner in which the numbers in any row (after the first two)
are formed from those above is clear from an examination of the
triangle itself. Specifically, with the exception of the end 1s, any
term is made up of the sum (or addition) of the two numbers
closest to it in the row above. However, for large values of the
power n, it is not necessary to record the complete triangle to the
row in question. Instead, it is quite possible to write down an
explicit expression for the term required. Thus, the $(m + 1)$th term
in the $(n + 1)$th row can be written as

$$C(m, n) = n!/[m!(n - m)!]$$

where $n!$ (called 'factorial n') is the shorthand notation for all the integers from 1 up to n multiplied together. For example, the third term in the seventh row should be $C(2, 6) = 6!/(2!4!)$, or equivalently $720/(2 \times 24) = 15$ which, as we see from the triangle above, it is.

In the coin-tossing exercise, this particular term should (if the association with probability is correct) count the number of ways that two heads and four tails can result from six tosses. In other words, it should be equal to the number of different ways in which I can write down a string of two Hs and four Ts. Let us check to see whether this is so.

Suppose that I could distinguish between all the Hs and the Ts in the string (say by writing them in different colors). In this case, all six letters in my strings are different so that I can place any one of the six in the first position. Following this, I have five remaining for the second position, four for the third, and so on; giving me 6! different strings in all. But in actuality I have no way of distinguishing between the two Hs or between the four Ts in each pattern, so that I have over-counted both — by a factor of 2! for the Hs (since $2! = 2$ is the number of ways of rearranging two distinguishable Hs) and a factor of 4! for the Ts (since $4! = 24$ is the number of ways of rearranging four distinguishable Ts). The answer to this particular coin-tossing problem is therefore just the number $C(2, 6) = 6!/(2!4!) = 15$ given by Pascal's triangle. The reasoning is obviously extendable to the general case of having $C(m, n)$ different ways in which m heads (or, equivalently, m tails) can occur in a sequence of n coin tosses.

Writing the $(n + 1)$th row of the Pascal triangle as

$$(H + T)^n = \sum_{m=0}^{n} C(m, n)H^{n-m}T^m$$

this equation can be turned into a statement about probabilities associated with the tossing of fair coins simply by dividing both sides by 2^n to give

$$(\tfrac{1}{2}H + \tfrac{1}{2}T)^n = \sum_{m=0}^{n} P(m, n)H^{n-m}T^m$$

in which

$$P(m, n) = n!/[m!(n - m)!2^n]$$

is the probability of tossing exactly m heads (or m tails) in n tosses.

Note that there is a 'line of symmetry' down the middle of Pascal's triangle which implies that $P(m, n) = P(n - m, n)$. For example, the probability of tossing three heads and seven tails in ten tosses of a fair coin must, by this symmetry, be exactly the same as the probability of tossing seven heads and three tails. The probability of each we now know to be

$$P(3, 10) = P(7, 10) = 10!/(7!3!2^{10}) = 0.117\,1875.$$

Not only does a formalism of this kind work for a fair coin, it also works with minimal modification for a biased one. For example, if a biased coin has a probability of falling heads over tails in the ratio of 3 to 2 for a single toss, then it is merely necessary to replace $(\frac{1}{2}H + \frac{1}{2}T)^n$ in the above by $[\frac{3}{5}H + \frac{2}{5}T]^n$. The probability of throwing three heads and seven tails in ten tosses using this biased coin then becomes

$$C(3, 10) \times (\tfrac{3}{5})^3 \times (\tfrac{2}{5})^7 = 0.042\,467 \ldots$$

where we have taken the appropriate term in the new binomial expansion. We note that this probability now differs from that for obtaining three tails and seven heads, which is

$$C(3, 10) \times (\tfrac{3}{5})^7 \times (\tfrac{2}{5})^3 = 0.214\,99 \ldots.$$

All this was known (to all but the statistically ill informed) by the end of the seventeenth century. In fact, there is evidence that the pattern of coefficients in Pascal's triangle was known to the Chinese as early as the fourteenth century. However, the explicit algebraic formalism $C(m, n) = n!/[m!(n - m)!]$ for general m and n seems first to have been given by Isaac Newton in letters to a colleague dated 13 June and 24 October 1676. Newton's proof has not survived, the oldest extant proof of this finding appearing in a book entitled *Ars Conjectandi*, by Jakob Bernoulli (1654–1705) which was published posthumously in 1713.

Bernoulli's book was a treatise on probability theory discussing, among other things, applications to insurance and to heredity. From our coin-tossing point of view, however, it also contained within it perhaps the most misunderstood of all notions concerning chance—the so-called 'law of large numbers'. In common parlance it says that in a large number of tosses of a fair coin one should get an approximately equal number of heads and tails. This seemingly innocent 'law' has been the downfall of many a gambler. Gamblers, you see, nearly always interpret it to mean that, if a fair coin falls dominantly heads in a sequence of tosses then, in later tosses, tails must become favored 'to even out the averages'. Otherwise, the argument goes, the number of heads and tails will not approach equality in the manner (they think) that the law of large numbers requires. This, of course, is all patently nonsense (coins cannot remember the past) and is an excellent recipe for losing wagers. It results not from any invalidity of the 'law of large numbers' but in a basic misunderstanding of what the law really says.

The law states that, as the number of tosses increases to larger and larger values, then the *ratio* of the number of heads, $m(H)$, to the number of tails, $m(T)$, gets ever closer to one. In other words, if we write

$$m(H)/m(T) = 1 + d(n)$$

where $d(n)$ is a function of the total number n of tosses, then this number $d(n)$ is small with respect to 1 (positive or negative) if n is large, and tends to get progressively smaller as $n = m(H) + m(T)$ increases. If we rewrite this equation in the form

$$m(H) = m(T) + d(n)m(T)$$

then we observe that the absolute difference between the number of heads and tails can be expressed

$$m(H) - m(T) = d(n)m(T)$$

which is the product of a large number, $m(T)$, close in value to $\frac{1}{2}n$, and a small number $d(n)$. Now the product of two numbers, one of which, $m(T)$, gets forever bigger as n increases, and one of which, $d(n)$, gets forever smaller, may either increase or decrease with n depending on the detailed forms of the numbers involved. It happens, as we shall see below, that for the case of coin tosses this product is known to increase indefinitely with increasing n. Thus, although the *ratio* $m(H)/m(T)$ of heads to tails does indeed approach ever closer to one, the difference between the number of heads and the number of tails $m(H) - m(T)$ actually gets forever larger. The gambler's expectation that $m(H)$ and $m(T)$ must eventually approach equality if the sequence is continued far enough is just not true. Put another way, if all the people in the world started a coin-tossing exercise, and each was allowed to stop only when they had tossed an equal number of heads and tails then, even though half of them would finish the project in just two tosses (one head and one tail), a very large number would still be playing after a lifetime of effort and with every expectation of playing on to eternity.

Part of the problem is that long sequences of tosses are almost certain to contain long runs of consecutive heads or tails. For example, in a baseball season of 162 games, even if the teams are all of exactly equal strength (in the sense of being equally likely to win or lose any individual game), then a run of seven or eight consecutive losses (or wins) is to be expected on statistical grounds alone. Since this contradicts most people's (and particularly sportswriters') ideas of randomness, these runs of wins or losses will almost certainly be 'explained' on some imagined psychological grounds (in terms of 'slump' for losses or 'momentum' for wins) even though it is more likely to be merely statistics at work in a mathematically anticipated, but almost universally unappreciated, fashion. Most people, if asked to write down a sequence of 162 Ws (for wins) and Ls (for losses) in what they perceive to be a random fashion (such as might occur for two equally matched teams), avoid such runs and almost invariably finish up with a much smaller difference between the total number of wins and losses than would actually be anticipated on firm mathematical grounds. This ingrained belief in the layman's interpretation of the 'law of large numbers' (or what he will more likely call the 'law of averages') is so persistent that it is almost always possible to distinguish popularly composed 'random' sequences from actual (e.g. tossed-penny) ones by these attributes alone.

The mathematics of probability theory was therefore on a firm footing by the early 1700s and was performing admirably in principle, but with one very serious practical limitation. It concerned the all-important number $C(m, n) = n!/[m!(n - m)!]$, the sheer size of which, when n progressed beyond, say, 30 or 40, became such that most calculations containing it could not be carried out without enormous computational effort. Indeed, for values of n larger than about 100, most numerical values for $C(m, n)$ — that is, all except those with very small values for m or for $n - m$ — just could not be handled at all. The essential difficulty was the value for $n!$ for larger values of n. Try calculating 20!, for example, using just pencil and paper and you will begin to appreciate the problem. What was needed to make probability theory useful in a general sense was an accurate approximation for $n!$ of a form that could be handled even to the largest values for n.

Such a formula was determined independently by the British mathematicians James Stirling (1692–1770) and Abraham de Moivre (1667–1754). Although Stirling is generally credited with the first proof, in 1730, and the formula carries his name to this day, it was known earlier to de Moivre and it was de Moivre who advanced its use for elucidating statistical problems, particularly those associated with Bernouilli's law of large numbers. Stirling's formula, which becomes increasingly accurate as n gets larger and larger, is

$$n! \simeq (2\pi n)^{1/2}(n/e)^n$$

where $\simeq$ indicates an approximate equality. Using this, de Moivre was able to establish a remarkably simple formula for $C(m, n)$ based upon the exponential function e^x. He showed in particular that, since $C(m, n)$ is symmetric about its largest value at $m = \frac{1}{2}n$, it can therefore be expressed most simply by introducing the variable $k = m - \frac{1}{2}n$ that measures 'distance' from the peak value $C(\text{max}) = C(\frac{1}{2}n, n)$ at $m = \frac{1}{2}n$. In this manner de Moivre was able to establish the form

$$C(m, n) = C(\text{max}) \exp (-2k^2/n).$$

The equation is not a true equality, of course, since it makes use of the Stirling approximation, but its quantitative accuracy is impressive, particularly for large values of n (which is just the regime that had earlier given computational problems). In figure 43 we show a plot of this exponential form for the case of $n = 100$. On the scale of the figure it is not possible to distinguish the approximation from the exact values. The major discrepancies occur way out in the wings of the plot (i.e. for k less than -20 or greater than $+20$) for which the curve, on the scale of figure 43, cannot be distinguished from the horizontal axis.

Since $C(\text{max})$ in the curve for $n = 100$ takes a value a little larger than 10^{29}, the immensity of the task of calculating exact values $C(m, 100)$ for all m can well be imagined. However, since we are usually concerned with probabilities that sum to unity (the sum, in this context, being essentially the area under the curve of figure 43) the actual value for $C(\text{max})$, which is the only 'tough' number now left in the exponential approximation of the problem, is of no consequence. Putting $x^2 = 2k^2/n$, and integrating to find the area under the curve, we obtain the variously named normalized distribution function, error function, or Gaussian

$$f(x) = (1/\sqrt{\pi}) \exp (-x^2).$$

De Moivre, who was actually born in France although he lived most of his life in England, first published the derivation of this formula in the second edition of his fundamental work *Doctrine of Chances* which appeared in 1738, although he had already made the basic result known to colleagues as early as 1733. What excited him most about the result was the insight it could give to Bernoulli's law of large numbers; a quantitative unravelment of that much misunderstood 'law of averages'. In this context, the

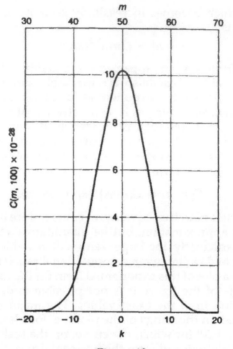

Figure 43

most significant property of the Gaussian function is the fact that it falls to a fraction $1/e^2$ (or about 0.1353, since e = 2.718 28 ...) of its peak value when $k^2 = n$ or, equivalently, when k is equal to plus or minus $\sqrt{n}$. It follows that $2\sqrt{n}$ measures, in some sense, a 'width' of the distribution—that range for which the probability is significantly non-zero. To be more specific, 95.44% of the area under the Gaussian function lies between these plus and minus $\sqrt{n}$ bounds. But even more importantly, as a function of the total number (n) of possible outcomes $C(m, n)$—that is, with m running from 1 to n, or k from $-\frac{1}{2}n$ to $+\frac{1}{2}n$—the *fraction* of m- or k-values that produces a probability 'significantly' different from zero is only of order $2\sqrt{n}/n$ or, equivalently, $2/\sqrt{n}$. Thus, as n increases, this fraction gets smaller and smaller, and the bell shape of the Gaussian function gets squeezed narrower and narrower, see figure 44.

Returning to the example of coin tosses, for one hundred tosses ($n = 100$) there is a 95.44% probability that the number of heads

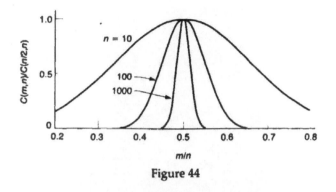

Figure 44

(or tails) obtained by experimentation lies between $0.4n$ and $0.6n$ (that is, between 40 and 60) as is apparent from figure 43. However, for a million tosses ($n = 10^6$) there is a 95.44% probability that the number of heads (or tails) lies between $0.499n$ and $0.501n$ (that is, between 499 000 and 501 000). Thus, whereas the spread in the *ratio* of heads to tails rapidly gets smaller as n increases, the actual number of m-values inside this region rapidly gets larger. This is the 'law of averages' as it should be understood. Put another way, as the number of tosses increases to large values, the probability that the ratio of the number of heads to tails is *approximately* equal to one approaches certainty. On the other hand, the probability that the number of heads and tails is *exactly* equal becomes forever smaller (specifically about 8% for $n = 100$ and 0.08% for $n = 1 000 000$). The seemingly paradoxical gulf between these two statements was therefore successfully bridged by de Moivre who was, with some justification, very pleased with the outcome, deeming it to be one of the 'hardest problems that can be posed on the subject of chance'.

In spite of de Moivre's enthusiasm, this discovery received little attention until Laplace began his study of probability theory in the 1770s. It was at this time that the first association of the 'error' function with accuracy of measurement in repeated observations was made. Being an astronomer as well as a mathematician, Laplace's interest related particularly to astronomical observations. If the sources of inaccuracy in such repeated measurements were deemed to be numerous and independent, then the problem was in essence a disguised form of the coin-tossing exercise (each individual source of error being associated with a head, say, for positive error and a tail for negative error). Thus followed a theory of 'random errors' in which de Moivre's

James Clerk Maxwell, 1831–1879. Courtesy of The Science
Museum, London.

exponential function $(1/\sqrt{\pi})\exp(-x^2)$ again appeared, now disposed symmetrically about the 'true' value for the quantity under
observation, and with a width that could be related to the
accuracy of the data in a precise mathematical fashion.

This association with measurement gave birth to the science of
'statistical analysis'—a mathematical method for analysing
observational data in a manner that enabled precise conclusions
to be obtained from empirical results. The theory was embel-
lished and worked out with great sophistication in the decades
following Laplace's early work, and was adapted for use in an
extensive domain of application from the physical to the social
sciences. Somewhat surprisingly, perhaps, this theory of random
errors first found focused application in the context of the social
and behavioral sciences, establishing the viability of mortality
and crime rates in different populations, for example, and adapt-
ing the findings for the production of insurance tables. Nevertheless, an explosion of activity in observational astronomy, at the

Ludwig E Boltzmann, 1844–1906. Courtesy of University of Vienna. Reproduced by permission of AIP Emilio Segrè Visual Archives.

beginning of the nineteenth century, proved to be a fertile ground for the use of the new techniques of statistical analysis and ensured that the methods would not be lost to the field of physical science.

Although the methods of statistical analysis in connection with interpretations of experimental data were increasingly accepted as the nineteenth century progressed, the seminal breakthrough (from the point of view of modern physics) using these concepts did not take place until the 1860s. It was at this time that the British physicist James Clerk Maxwell (1831–1879) and his Austrian-born contemporary Ludwig E Boltzmann (1844–1906) independently recognized that these probability functions were just the tools necessary to bridge the yawning gap between the microscopic and macroscopic descriptions of the properties of matter. Up until this time there were two quite separate theoretical sciences that dealt with the properties of matter (i.e. gases, liquids and solids). The first was classical mechanics—in terms of which gases, for example, could be pictured as atom-sized 'billiard balls' traveling at great speeds in a somewhat chaotic manner, forever colliding with each other and with the walls of the container. Although Newtonian equations of motion could in principle be written for this situation, they could only be solved if very few atoms were present, in which case the language of interpretation contained such notions as time, mass, velocity, energy, force and so on. The practical problem that rendered this approach essentially useless for real gases was the sheer number (up to 10^{23}) of atoms or molecules that are commonly present in bulk samples, together with the fact that the detailed forces involved in collisions (particularly with the microscopically 'rough' walls of any container) were poorly defined. The resulting Newtonian problem, even if it could be spelled out in some idealistically simple situation, was (and still is) far beyond any computational powers to solve.

It was for precisely this reason that the dominant scientific language used at that time for bulk materials was completely different, and based on a few directly measurable macroscopic quantities. The language was 'thermodynamics', a science governing the general laws of heat effects and based on such concepts as temperature, entropy, pressure, volume, heat capacity, thermal conductivity, and so on. This macroscopic description had been finalized in the mid-nineteenth century by such pioneers as the German physicist Rudolf Clausius (1822–1888) and his British contemporary William Thomson (later Lord Kelvin 1824–1907). In thermodynamics, heat is recognized as a form of energy and is related, by self-consistent relationships, to other large-scale energies of the system under investigation. These energies, and other related thermodynamic properties, generally involve no special assumptions about the structure of matter and are suggested, more or less, by our sense perceptions. Most importantly, they can be directly measured and enable the theory, which predicts relationships between them, to be tested. At no point, however, is any contact with the microscopic world explicitly developed although it was always recognized that the measurable properties that completely determine the thermodynamic description must really be averages over time of an extremely large number of microscopic variables. For example, in a gas, the thermodynamic quantity pressure, which can be perceived by our senses, is actually the result of the impacts of atoms or molecules against the walls of the containing vessel. The notion of pressure, however, could be experienced, measured, and used long before physicists had any idea of how to derive a mathematical expression for it in terms of the motion of atoms and molecules. Thermodynamic formulae therefore related macroscopic quantities one to another but nowhere made contact with any of the ideas of classical (or Newtonian) mechanics.

The problem of determining the motions of exceedingly large numbers of colliding bodies first came to Maxwell's attention as early as the 1850s, but he dismissed it as hopelessly intractable. Then, in 1859, he chanced to read a new paper by Clausius, who himself had made his first venture into the kinetic theory of gases two years earlier. In order to attack the problem of deriving an expression for the average path length d of a molecule between collisions, Clausius first imagined only one molecule to be moving, with the remainder fixed in an essentially lattice-like framework; but later he improved this to allow all molecules to move. He found a relationship

$$1/d = \tfrac{4}{3}\pi s^2 N$$

where s is the molecular diameter and N is the number of molecules per unit volume. The primary weakness in the calculation was an assumption that all the molecules had the same velocity. Although Clausius recognized that the velocities would, in actuality, be spread in some fashion, he had no way of estimating the actual form of the distribution. Inspired by this work, Maxwell began to ponder the problem of the velocity distribution itself for a gas under uniform pressure. He reasoned as follows.

Let the components of molecular velocity v along the three Cartesian axes be v_x, v_y, v_z such that

$$v^2 = v_x^2 + v_y^2 + v_z^2.$$

If the x, y, and z components of the motion are independent, then the number of gas molecules with velocities between v_x and $v_x + dv_x$, v_y and $v_y + dv_y$ and v_z and $v_z + dv_z$ (where the dvs are small differential increments) can be written as

$$dN = Nf(v_x)f(v_y)f(v_z)dv_x dv_y dv_z$$

where f is an as yet unspecified function, and there are N molecules in total. Since the axes are arbitrary, dN can depend only upon v so that

$$f(v_x)f(v_y)f(v_z) = \phi(v_x^2 + v_y^2 + v_z^2)$$

where ϕ is some other function. It is now easy to show that the only possible solution has to be of the form

$$d(N_x) \sim \exp{(-v^2/a^2)}dv_x$$

where a is a quantity with the dimension of velocity. The number

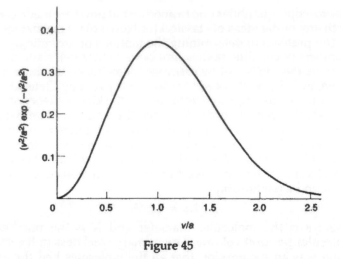

Figure 45

of molecules dN, summed over all directions, with velocities
between v and $v + dv$ must therefore vary as

$$dN \sim v^2 \exp(-v^2/a^2)dv$$

where the extra v^2 has arisen from expressing the elementary
'volume' dv_xdv_ydv_z in polar coordinates (i.e. the volume of a shell
at 'radius' v is $4\pi v^2 dv$). This Maxwell velocity distribution curve
is sketched in figure 45. We see, interestingly enough, that the
fraction of molecules with both extremely small and extremely
large velocities approaches zero, while the peak (and average)
velocities depend only upon 'a', which leads to a suspicion that 'a'
must be related to temperature. The distribution was experimentally verified some years later, but the fundamental assumption
(that the x, y, and z components of velocity are independently
distributed) was slow to be accepted. In particular, Maxwell
simply asserted that the distribution function must satisfy certain
properties. Nothing was said about the properties of molecules or
their collisions, which would presumably be responsible for
bringing about the distribution. Also, from our point of view,
there is as yet no obvious connection between Maxwell's method
of obtaining the error function and its Laplacian association with
the statistics of coin-tossing.

Later Maxwell (reacting to the criticisms) offered another derivation tied to the realization that, since molecular collisions occur
with such tremendous frequency then, for a bulk system in

equilibrium, the number of collisions involving two atoms with approaching velocities v_1, v_2 and departing velocities v_3, v_4, must be essentially equal to the number of collisions of the reverse kind. He then showed that this equality could prevail only if his velocity distribution law again held. The implication was that, once molecules attained such a 'Maxwellian' distribution, collisions would not disturb it. But this proof was also unsatisfactory since the result is strictly true only as an average over long periods of time. Nevertheless, in spite of these weaknesses, Maxwell pressed on to successfully apply his distribution function to evaluate such macroscopic quantities as viscosity, diffusion and heat conduction and, with such calculations, marked the beginning of a new epoch in physics—the prediction of macroscopic properties from a microscopic starting point.

Boltzmann's first major achievement, in 1868, was to extend Maxwell's distribution law to the case where an external field (like gravity) was present. For such a case the probability distribution in space is no longer constant so that the distribution function must now explicitly contain both spatial and velocity (or equivalently momentum) coordinates. But, as with Maxwell, Boltzmann did not succeed in proving his relationships except in some very special cases. Nevertheless, it was he who successfully first introduced the probability arguments that make it much clearer why a system in equilibrium should obey laws involving the Gaussian function and be intimately related to statistics (of which the coin-flipping kind are particular examples).

Over the next few decades Maxwell's work was extended in a number of ways, particularly to elucidate the role played by temperature and to apply the statistical reasoning to systems other than gases. Finally, by the end of the century, the statistical technique was actually in rather good shape. The final task was to put all the mathematical principles into a form that was independent of the specifics of any particular microscopic model. This final step was primarily the work of the American J Willard Gibbs (1839–1903). Looked at from today's world of quantum physics, classical problems (like that of a gas of molecules) are much more difficult than their quantum counterparts because quantum phenomena involve discrete events (like coin tosses) that are easy to count, while classical phenomena involve continuous motions and calculus. The luxury of discrete counting (at least in an overt form) was not available to Boltzmann and Gibbs and it is hard to believe that they derived their beautiful theoretical structure without it. It leaves us all the more impressed that they did.

In order to demonstrate the detailed manner in which a bridge between the microscopic world of atoms and the thermodynamic world of temperature and entropy can be constructed, we shall take full advantage of the now-recognized quantum aspects of nature and set up the problem in the context of magnetism. Here we shall use as our basic microscopic unit the atomic magnet (or magnetic 'moment' to give it its proper name) that can be associated with each and every magnetic atom in a solid.

Consider a solid made up of $n \sim 10^{22}$ or 10^{23} magnetic atoms, each of which possesses a microscopic magnetic moment (or 'spin') s that can either point in an 'upward' direction ($+s$) or a 'downward' direction ($-s$), up and down being determined by crystalline constraints that need not worry us. If all n spins are 'up' or all are 'down', then we have a bulk ferromagnet with a total macroscopic spin of $+ns$ or $-ns$ respectively. This can arise spontaneously if there are strong interactions between the atomic spins which tend to favor a parallel alignment, but in our simplest of all models we shall neglect such forces (an assumption that is quite realistic for some magnetic materials).

Associating a head with an 'up-spin' and a tail with a 'down-spin' the problem of determining the macroscopic properties of our n spins now reduces exactly to a coin-tossing exercise. Thus, in the absence of forces between atomic moments the probability of having $\frac{1}{2}n + k$ spins 'up' and $\frac{1}{2}n - k$ spins 'down' is just our old binomial friend $P(m, n) = C(m, n)/2^n$ where $m = \frac{1}{2}n + k$ or $m = \frac{1}{2}n - k$ (it doesn't matter which). Making use of Stirling's formula this can also be written as

$$P(m, n) = P(\tfrac{1}{2}n, n) \exp(-2k^2/n)$$

or, expressed solely in terms of k, as

$$P(k, n) = P(0, n) \exp(-2k^2/n).$$

In the language of statistical mechanics, which is the 'bridge mathematics' that we are developing, $C(m, n)$ or equivalently $C(k, n)$ is called the 'degeneracy' or 'number of accessible states' of the configuration 'k' with $m = \frac{1}{2}n + k$ spins 'up'.

Suppose now that we have two such systems n_1, k_1 and n_2, k_2 and bring them into thermal contact. By this we mean that they are allowed to exchange energy but only in a manner which obeys the law of conservation of total energy. To introduce energy into the problem we apply a small magnetic field H. For the configuration k, with $2k$ more spins 'up' than 'down' and a consequent total magnetic 'moment' of $M = 2ks$, physics tells us that the associated magnetic energy is

$$MH = 2ksH.$$

It follows that energy can only be exchanged in a manner that conserves the quantity

$$U = 2(k_1 + k_2)sH$$

or what is that same thing for constant field H, conserves $k_1 + k_2$.

What will happen to k_1 and k_2 as the two systems are allowed to exchange energy (i.e. come to thermal equilibrium) in this fashion? Well, the total number of accessible states of the combined system is

$$C(K, N) = \sum_{k_1} C(k_1, n_1)C(K - k_1, n_2)$$

where we have written $K = k_1 + k_2$, $N = n_1 + n_2$ and included the required energy restriction.

The above sum, using Stirling's approximation, is a sum over pairs of exponential products like

$$\exp(-2k_1^2/n_1) \exp[-2(K - k_1)^2/n_2].$$

A sum of products of this kind peaks *extremely* sharply at the maximum value, which is the value for which

$$k_1/n_1 = (K - k_1)/n_2 = k_2/n_2 = K/N.$$

An indication of how sharp this peak is can be gleaned from figure 46 in which we plot the product of $\exp(-x^2)$ and $\exp[-(50 - x)^2]$ as a function of x and observe the peak at $x = 50 - x$, or $x = 25$. For our magnetic example, with n_1 and n_2 being numbers like 10^{22}, the function is so peaked that fractional deviations of k_1/n_1 and k_2/n_2 from their equilibrium values by an amount even as tiny as 10^{-10} are unlikely to occur in the lifetime of the universe.

An exactly parallel argument can be given replacing $2k_1$ and $2k_2$ by their energy equivalents U_1/sH and U_2/sH. In this case we arrive at the thermal equilibrium condition

$$U_1/n_1s^2H^2 = U_2/n_2s^2H^2 = U/Ns^2H^2$$

implying that, for any particular system, the quantity U/ns^2H^2 must be some measure of temperature. This ratio could, in fact, be used to define any temperature scale we like. However, if we wish it to be the same scale as that developed historically in connection with thermodynamics (with T measured in degrees Kelvin from a value of $T = 0$ at absolute zero) then the required relationship is

$$U/ns^2H^2 = -1/\kappa T$$

in which κ is a constant of proportionality called Boltzmann's

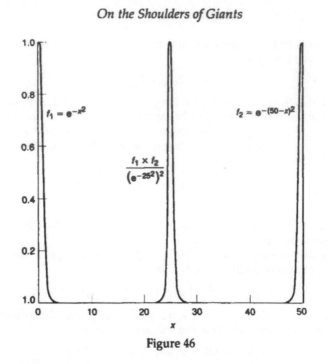

Figure 46

constant. It follows, therefore, that the equilibrium thermal energy of n thermally interacting spins in an applied field H and at a temperature T is

$$U(T) = -ns^2H^2/\kappa T.$$

Thus, we say that two systems are in *thermal equilibrium* with each other when the combined system is in its most probable configuration; that is the configuration with the greatest number of accessible states (or degeneracy). The values for degeneracies are extremely large numbers and it is far more convenient to work with their natural logarithm. This logarithm, it turns out, is just the entropy of the system as defined by thermodynamics. Next to temperature, entropy is probably the most important concept in all of thermodynamics. It is defined such that a small increase dU in energy at constant temperature leads to a small increase in entropy according to the formula

$$dU = \kappa TdS.$$

Statistically it is now easy to see that S should be the natural logarithm of $C(K, N)$ since this degeneracy, written in terms of $U = 2KsH$ is just

$$C(K, N) = C(0, N) \exp \left[-U^2/2s^2H^2N\right]$$

the natural logarithm (ln) of which is

$$S = \ln \left[C(0, N)\right] - U^2/2s^2H^2N$$

and which, upon differentiation with respect to U, gives

$$dS = -UdU/Ns^2H^2 = dU/\kappa T$$

where we have made use of our earlier definition of T. Entropy is a dimensionless number that measures the degree of randomness in the system. Consequently, the more states that are accessible, the greater is the degree of the randomness and the larger is the entropy.

As a reward for all this tedious mathematics we are now in a position to derive what is undoubtedly the most famous result in all of statistical mechanics. It is called the Boltzmann distribution law, and answers the question (for our particular magnetic example) 'What is the probability in a given observation that, for spins in equilibrium in a magnetic field H at temperature T, a particular spin will be "up" or "down"? This ratio, which we may call $P(\text{up})/P(\text{down})$, is given by

$$P(K, N)/P(K - 1, N)$$

and takes the explicit form (remembering that $e^a/e^b = e^{a-b}$)

$$\exp \{(2/N) \left[(K - 1)^2 - K^2\right]\} = \exp \left[(2 - 4K)/N\right].$$

Neglecting 2 in comparison with $4K$ (since K, like N, is a number of immense magnitude) and making use of the energy relationship

$$U = 2KsH = -Ns^2H^2/\kappa T$$

now enables us to derive the delightfully simple finding

$$P(\text{up})/P(\text{down}) = \exp \left(-2sH/\kappa T\right).$$

Since the energy of an individual up state is $E(\text{up}) = +sH$ and the energy of an individual down state is $E(\text{down}) = -sH$, the above equation can now be written in its final form

$$\frac{P(\text{up})}{P(\text{down})} = \frac{\exp \left[-E(\text{up})/\kappa T\right]}{\exp \left[-E(\text{down})/\kappa T\right]}.$$

In a more general context this statistical mechanical 'rule' states that, if a large assembly of independent systems is in thermal equilibrium at a temperature T, and each system can exist in quantum energy states E_i ($i = 1, 2, 3, ..$), then the probability that any *particular* system will be in energy state E_i is simply proportional to the 'Boltzmann factor' exp $(-E_i/\kappa T)$.

As an example of the use of the Boltzmann factor we may now return to Maxwell's original problem concerning the velocity distribution of molecules in gases. Since the energy E of a molecule with velocity v is its kinetic energy $\frac{1}{2}mv^2$, where m is its mass, the Boltzmann factor (now, for a classical as opposed to a quantum problem, converted from a discrete sum to a continuous function) is exp$(-mv^2/2\kappa T)$. Since the total number of molecules dN with velocities between v and $v + dv$ is proportional to $4\pi\, v^2 dv$, the probability distribution for velocity must be

$$dN \sim v^2 \exp\, (-mv^2/2\kappa T)dv.$$

Compared with Maxwell's finding as given earlier, namely

$$dN \sim v^2 \exp\, (-v^2/a^2)dv$$

the earlier suspicion that the parameter 'a' should be related to temperature is confirmed in the form $a^2 = 2\kappa T/m$. In addition, the problem is now no more difficult if we imagine the molecules to be subjected to an external field (say gravitation). This just adds to the energy formula a (potential energy) term $-mgx$, where g is the gravitational acceleration and x is the vertical distance. Putting $E = \frac{1}{2}mv^2 - mgx$, statistical probabilities can now be pursued using the Boltzmann factor just as before, but where now the 'phase space' of allowed configurations must include (classically) all six dimensions, x, y, z, v_x, v_y, v_z.

This 'Maxwell–Boltzmann' statistical bridge between the microscopic and macroscopic world is now firmly entrenched as a foundation stone of modern physics. However, it is now recognized that the Maxwell–Boltzmann formalism for statistical mechanics is not the only such bridge that can be built. It assumes that the individual microscopic 'systems' that we are 'counting' can, at least in principle, be distinguished one from another and that each, independently, can occupy any of the allowed configurations. It is now known that for some quantum particles and more esoteric 'quasiparticles' this is not the case, and forms of counting slightly different in detail from that of coin-tossing must be used for them. The two most common other bridges from the microscopic to the macroscopic world are known as Bose–Einstein statistics and Fermi–Dirac statistics, respectively, each named after pairs of illustrious physicists of the twentieth century.

Nevertheless, the coin-tossing bridge remains valid for most situations involving atoms or molecules as individual systems.

Using this bridge, it has been possible to derive atomic models for a whole host of properties of gases, liquids and solids, pure materials, alloys and mixtures. Most experiments probe macroscopic properties as a function of temperature or pressure, while most fundamental theories are set up for elementary interactions between individual microscopic constituents. A test of such theories therefore usually involves the incorporation of a statistical mechanical bridge. As a result, statistical mechanical calculations of one form or another pervade the research literature of theoretical chemistry and physics today.

This is true not only for calculations concerning the thermal equilibrium state. Since the central peaks in figures like figure 46 are not infinitely sharp for real systems, but do have widths which are amenable to calculation, statistical mechanics can also probe fluctuations about thermal equilibrium. The most familiar property that receives explanation in this manner is the blue color of the sky which is due to light scattering from molecular density fluctuations in the upper atmosphere. However, many other 'fluctuation' phenomena in physics, particularly those involving instabilities like phase separation in liquids and order-to-disorder phase changes in alloys and magnets, can also be explained with great accuracy using these same methods. And all this from a few questions concerning the tossing of a couple of coins—by way of a few broad shoulders along the way!

10

Topology: from the Bridges of Königsberg to Polymers

Geometry, to Euclid, was a precise method of deducing relationships between angles, lengths, areas and volumes in planar or
solid figures. To the early geometers (and indeed to many
modern ones such as architects, bricklayers and carpenters) it
matters a great deal whether a triangle is right-angled or not, or
whether a quadrilateral is a square or a rectangle. It also matters
whether such an object is large or small—it is, for example,
important to know whether a house under construction is
planned for use by people or by dolls! Size and shape quite rightly
hold a respected place in the world of measurement. They are,
one might say, an essential element of what we can perhaps refer
to as 'metric' geometry—the geometry of 'measure'.

What other kind of geometry could possibly be of value? So
ingrained is the association of geometry with size and measure
that my dictionary actually defines the word 'geometry' as the
'science of properties and relations of *magnitudes* in space', seemingly precluding any more subtle concept. And yet there is
something in geometry that goes beyond the relationship of
magnitudes. In what sense, for example, do *all* triangles differ
from *all* quadrilaterals regardless of size and shape? They obviously do; and we can easily quantify this difference by simply
referring to the number of sides and corners.

There is, therefore, without any doubt, something that
uniquely differentiates between triangles and quadrilaterals
regardless of angular measure, size, or even whether the sides are
straight. It is at this deeper level of discrimination between
geometric objects that we meet the science of topology. To the
topologist, only the properties of objects that are retained during

any continuous deformation of space are worthy of note. It has been said that metric geometers measure only the transitory aspects of geometrical objects while topologists measure their ultimate essence (their very souls if you will).

To the metric geometer, squashing a ball represents change — it changes it from a sphere to an 'oblate spheroid', to be formal about it. To the topologist, on the other hand, squashing a ball represents constancy — the ball, squashed or otherwise, is a single closed surface with an unambiguous outside and inside. No matter how continuously distorted, its 'soul' remains intact. It will never, for example, become a doughnut. A doughnut, at least according to topology if not always at the doughnut shop, is a ring-shaped object with a hole through the middle. No amount of stretching or twisting can ever turn it into a ball; it possesses a different soul. A doughnut can, however, if you think about it, be turned into a coffee mug, from which follows perhaps the best-known humorous definition of a topologist, namely as 'someone who cannot tell the difference between a coffee cup and a doughnut'.

Topology is not a familiar word to the average layperson. In fact, many smaller dictionaries avoid mentioning the word altogether. Perhaps an adequate definition would be 'the study of those geometrical properties that are unaffected by changes of shape or size'. The size- and shape-modifying operations referred to in this definition would most usually be those of bending, stretching and twisting, and indeed these three operations are quite often referred to as 'topological transformations'. The properties that define a geometric object from a topological point of view can therefore be thought of as 'topological invariants'; that is, numerical properties that remain unchanged under any sequence of topological transformations. Two objects which are topologically equivalent are said to be homeomorphs, and it's a pretty good bet that your dictionary omits that one too! However, next time you sit down with your coffee and ring-doughnut, at least you will now know how to slip it into the conversation.

In spite of their general unfamiliarity with the word, laypersons probably unwittingly confront topological problems more often than they realize. Who, for example, has never seen a puzzle like that of figure 47, where the puzzler is asked to take a pencil and, without removing it from the paper, draw a continuous line (starting and finishing in any convenient location) that intersects each and every line segment of the figure once only? Puzzles of this kind are, of course, often chosen so that the task demanded is impossible to complete. In that way the puzzler can

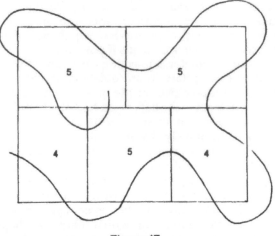

Figure 47

while away as many hours as patience allows trying to complete
an impossible assignment.

What is required to short-circuit all the trial-and-error is a
topological argument. Clearly the problem remains valid regardless of how much the figure is stretched or twisted, so that we
need to find some properties of the adjoined areas that are
independent of such distortions. One obvious one is the number
of separate line segments that together define the perimeter of
each area. For the pattern in figure 47, three of the areas have five
such line segments and the other two have four, as indicated in
the figure by the numbers placed inside the areas in question.
Now any line starting outside an area with an odd number of line
segments must finish up inside and vice versa. It follows that any
set of adjoined areas with more than two odd-numbered constituents cannot possibly be traversed by a line in the manner
required since, although the line can start and finish inside an
area, only two odd-numbered areas can be accommodated in this
manner. Any others will necessarily have sides that remain
uncrossed. Since the pattern of figure 47 has three odd-numbered
areas, no continuous line can be drawn to intersect each side once
only. Thus, using an almost trivially simple topological argument, we have proved that the problem as posed has no solution;
and isn't that better than countless doodling trials?

It is difficult to pinpoint exactly the 'birth' of the mathematical
discipline now known as topology. Many would cite the work of

Leonhard Euler, 1707–1783. Reproduced by permission of Mary Evans Picture Library.

Leibniz (1646–1716) who, in his book *Characteristica Geometrica*, which was published in 1679, did make some effort to study the properties of geometrical figures in terms of what we would today call their topological rather than their metric parameters. He referred to the subject, such as it then was, as *'analysis situs'* (the analysis of position). He wrote that, in addition to the coordinate representation of figures 'we need a different analysis ... which defines the position (*situs*) as algebra defines quantity'. Leibniz tried to interest others in the idea (particularly the Dutch astronomer, physicist and mathematician Christian Huygens (1629–1695), who is perhaps best remembered for his work in optics) but without much success. In fact, in the following two hundred years or so, very little else appeared on the topic of *'analysis situs'*, with one very notable exception. This was the signal work of Leonhard Euler who, among his many other fields of mathematical interest, found the time to set out some relationships that exist between what would now be called the topologi-

cal features of planar and solid figures. These topological features, for any particular geometric object, are no more than the
number of areas (or *faces*) *f*, the number of lines (or *edges*) *e* that
define the areas, and the number of points (or *vertices*) *v* at which
the edges intersect. In searching for relationships between *f*, *e*,
and *v*, Euler was obviously concerned with properties that would
not change with stretching, bending and twisting. Thus,
although he certainly never used the word topology, it is these
Eulerian relationships that many would cite as the earliest significant quantitative accomplishments in what eventually would
become the field of topology. This being the case, we shall pursue
them in a little more detail below.

Consider, for example, the somewhat arbitrarily drawn net of
connected polygons depicted by the full lines in figure 48(*a*). It
contains 12 vertices, 16 lines (or edges) and 5 enclosed areas (or
faces). Thus, in our notation, we have $v = 12$, $e = 16$ and $f = 5$, and
it is quite self-evident that this will remain the same no matter
how we continuously distort the figure. Suppose now that we
focus upon the combination

$$I(2) = v + f - e$$

where the 2 in parentheses denotes that we are discussing a
planar (or two-dimensional) figure, as we alter the topology of the

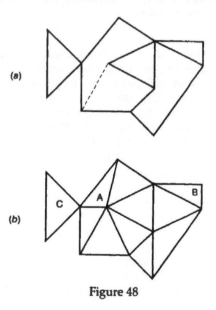

(*a*)

(*b*)

Figure 48

net. First we note that $I(2) = 1$ for our starting net. Let us now draw an arbitrary face diagonal, such as that shown dashed in the figure. This maintains the number of vertices (i.e. v remains unchanged) while increasing each of e and f by 1. It therefore leaves $I(2)$ still equal to 1. This procedure can be repeated over and over again, each time leaving $I(2) = 1$, until the network consists wholly of triangles, as shown in figure 48(b) with $v = 12$, $e = 22$, and $f = 11$.

Now let us consider removing triangular areas one at a time from the boundary. This can be accomplished (quite generally) in only three distinct ways, exemplified respectively by the triangles labeled A, B, and C in figure 48(b). For each case, the number of faces is decreased by 1 (i.e. $f \rightarrow f - 1$), but the three differ with respect to their changes in v and e as follows:

for triangle A, $v \rightarrow v$, $e \rightarrow e - 1$;
for triangle B, $v \rightarrow v - 1$, $e \rightarrow e - 2$;
for triangle C, $v \rightarrow v - 2$, $e \rightarrow e - 3$.

In each case we easily verify that $v - e \rightarrow v - e + 1$ and therefore, since $f \rightarrow f - 1$ in all cases, $I(2) = (v - e) + f = 1$ remains unchanged during the process. The endpoint of the entire triangle-removing exercise is a single triangle with $v = 3$, $e = 3$, $f = 1$ for which (now no longer surprisingly) $I(2)$ is still equal to 1. Reversing the entire procedure, it follows that *any* connected network of polygons in two dimensions must always obey the relationship

$$I(2) = v + f - e = 1.$$

Even adding dangling endlines to the net (with $v \rightarrow v + 1$, $e \rightarrow e + 1$) does not affect the constancy of $I(2)$ which therefore becomes a topological invariant for any two-dimensional connected network whatsoever.

Using his growing familiarity with topological concepts, Euler liked to challenge his colleagues with problems that require topological solutions. The most famous of these is known as the 'Bridges of Königsberg' problem. In Euler's day, Königsberg was in Prussia. After the First World War it found itself in the USSR disguised as the distinctly non-Germanically sounding city of Kaliningrad. Where it will be, or by what name it will be known, by the time you read this I do not know. Fortunately that will not affect our story. The Königsberg that Euler knew had its town center on an island that was connected to the surrounding 'suburbs' by seven bridges over the Pregel river, as shown in figure 49. The problem concerned the challenge of starting a walk at any desired point and proceeding to any other by crossing each

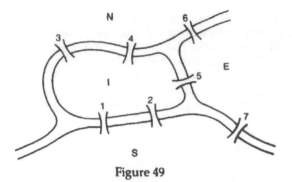

Figure 49

bridge only once. Indeed, was such a walk possible? The problem
is obviously related in spirit to that of figure 47 and is, equally
evidently, well posed even if the map is distorted by any continuous deformation. Its solution is therefore topological, and Euler
dealt with it in the following manner.

The river, as the figure shows, divides the town into four
regions which we label I (for Island) and N, S, E (for the North,
South, and East suburbs, respectively). If we now represent these
regions by points corresponding roughly to their geographic
locations, and the seven bridges, denoted by the integers 1, 2, 3,
..., 7 in figure 49, by corresponding lines joining the points I, S, N,
and E, then the two representations of Königsberg can be made
identical as far as topological travel via the bridges is concerned
(see figure 50(*a*)). The essential point is that if each bridge is to be
crossed once, then there must be an even number of edges
meeting at each vertex of the representation in figure 50(*a*). This is
because for each edge (or bridge) used to reach a particular vertex
(or suburb), there must be another edge (or bridge) available for
use in continuing the walk. As the question was posed, however,
there may be two exceptions, namely the starting and finishing
vertices. But we see immediately from figure 50(*a*) that there are
three vertices at which an odd number of edges meet. It follows
that the walk envisaged by Euler is not possible no matter where
one chooses to start or finish. More accurately I should say 'was'
not possible for, since Euler's day, two more bridges have been
built across the Pregel in the positions designated by lines 8 and 9
in figure 50(*b*), i.e. one (8) across the previously unbridged most
western branch of the river, and the other (9) paralleling bridge 7
between the southern and eastern suburbs. Now, with only two
odd-numbered vertices (I and S) remaining, Euler's journey can

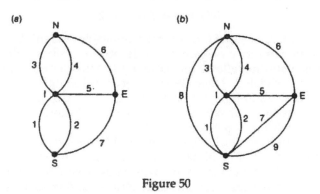

Figure 50

at last be taken—so long as one starts on the Island (I) and finishes in the southern suburb (S) or vice versa.

Having thus far concerned ourselves with topological problems in two dimensions, it is a natural extension now to progress into three dimensions, and it is in this context that Euler's most famous topological formula is found. It arises as the result of a search for a three-dimensional counterpart of the two-dimensional topological invariant formula $I(2) = v + f - e = 1$. At the outset, of course, there is no guarantee that such an invariant even exists; but a surprisingly simple argument shows that, at least in a limited capacity, it does.

Consider an arbitrarily complex polyhedron (three-dimensional by definition) characterized on its surface by polygonal faces, edges, and vertices in numbers f, e, and v, respectively. The limitation referred to above concerns the fact that not every solid body with faces, edges and corners is a polyhedron, a point that we shall expand upon below. But, returning to our polyhedron, and making use of the fact that in any topological investigation we may arbitrarily deform the object in question, we first remove one face, after which the closed solid becomes open. It is now possible to deform this remaining object into a plane polygonal net by stretching open the hole (i.e. the missing face) to the point where one can lay down the whole figure onto a plane with the missing face as the periphery or outside boundary. For this now-two-dimensional net we know, from our earlier investigation, that $I(2) = v + f - e = 1$. If we now reverse the deformation process to include the final step of putting back the face that was originally removed, this final step affects only the topological number f (with $f \rightarrow f + 1$), from which it follows immediately that, for all three-dimensional polyhedra,

$$I(3) = v + f - e = 2.$$

This particular proof (one of many available) is, I think you will agree, an extremely elegant example of the flexibility that continuous distortion allows us, reducing the topological argument almost to triviality. Nevertheless, for the doubters, perhaps a quick verification for a simple example is in order. Consider the cube. It has eight corners ($v = 8$), six faces ($f = 6$) and twelve edges ($e = 12$) with $I(3) = 8 + 6 - 12 = 2$.

Although Euler, in 1758, was the first to publish a proof of the existence of the three-dimensional topological invariant $I(3) = v + f - e = 2$, the actual result was known earlier, certainly to Descartes in the early-seventeenth century and possibly even to the Ancient Greeks. We note, in particular, that it is not at all necessary that the polyhedron be without 'dents' (or to be 'convex' as the geometers would put it). On the other hand, as mentioned at the outset, it must also be recognized that not all solid objects are polyhedra in the topological sense of being wholly contained by 'unbroken' faces, planar or otherwise. In order to make clear what we mean by this statement, take a look at the solid depicted in figure 51. If we count its vertices, faces and edges we obtain the numbers $v = 16, f = 11, e = 24$, and hence

$$I(3) = 16 + 11 - 24 = 3.$$

Had the object consisted of two separate blocks, one sitting on top of the other, all would have been well. It is the common 'face' of contact which, in the completely conjoined solid, has 'disappeared' that has created the problem. In particular, it gives rise to a face (the larger of the upper horizontal faces in the figure) that has a hole in it. This face is not a conventional (that is, closed) polygon and thereby disqualifies the solid in figure 51 from being a topological polyhedron, no matter how it is continuously distorted.

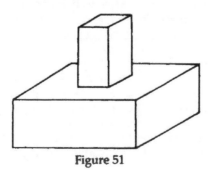

Figure 51

Using the Eulerian result $I(3) = 2$ for three-dimensional poly-hedra it becomes quite straightforward (although we shall not develop the detailed argument here) to establish that there can be only five regular solids—namely, the

tetrahedron: $v = 4, f = 4, e = 6$
octahedron: $v = 6, f = 8, e = 12$
icosahedron: $v = 12, f = 20, e = 30$

all with triangular faces; the

cube: $v = 8, f = 6, e = 12$

with square faces; and finally the

dodecahedron: $v = 20, f = 12, e = 30$

with pentagonal faces. They are shown in figure 1 of chapter 2.

To this point, however, we have concerned ourselves only with the topological invariant for a single polyhedron. In two dimen-sions, as we have seen, the result $I(2) = 1$ for a single polygon remains valid when extended to a whole planar 'froth' of connec-ted polygons. It is now of interest to inquire whether the polyhed-ral value $I(3) = 2$ is also valid when extended in an analogous three-dimensional fashion. Euler also considered this situation and found that such is not the case, an extension to a three-dimensional polygonal 'froth' requiring a little extra compu-tation. We may quickly get some inkling that this must be the case by considering a 'two-cell' froth made up of two face-joined tetrahedra for which we easily calculate $I(3) = 5 + 7 - 9 = 3$. The full extension to an arbitrary three-dimensional froth of polyhed-ral cells is not difficult (although we shall omit the details) and leads to the generalized finding of $I(3) = v + f - e = N + 1$, where N is the number of cells in the froth.

Given these extremely simple findings it is easy to get the impression that all topological issues of this kind are not too difficult to solve. Nothing could be further from the truth. In fact, there have been problems even in two-dimensional net systems that have taken generations, together with the most extraordi-nary of methods, to solve. The most infamous of these is un-doubtedly the 'four-color' problem. This concerns the number of colors needed to color a map, by which we mean any conceivable map, subject only to the restriction that any two areas (or 'countries' if you prefer) that possess a common boundary line should have a different color. The problem appears first to have been posed by a recently graduated London University student named Francis Guthrie in 1852. The Eulerian result $I(2) = 1$ is certainly not without value in probing the problem and, using it, a proof that at least five colors are sufficient is now a standard exercise in text-book topology. The problem, in 1852, was that Francis Guthrie could not seem to find any actual map that did, in fact, require all five. Try as he would, no matter how cleverly he attempted to devise otherwise, four colors seemed always to suffice. So perplexed was he by this fact that he postulated that four colors just might *always* be sufficient—and thus was born the 'four-color' problem.

The story now moves on some twenty years or more to 1878, in which year the eminent and prolific British mathematician Arthur Cayley (1821–1895), Professor of Mathematics at Cambridge, finding himself unable to settle the question, decided to set it before the London Mathematical Society. Shortly thereafter, one of the members of the Society named Arthur Kempe (who was a lawyer by profession) came closer to proving the four-color conjecture than any unassisted human has to this day. And I word this rather carefully because the great notoriety of the problem today rests upon the fact that a complete proof has so far been achieved only with the assistance of vast computational efforts by modern electronic computers. But to go back to Arthur Kempe—it was he who established the method which has been used for attacking the problem ever since he published his original efforts in 1880.

Kempe first used the basic Eulerian topological theory to establish that every map on a plane surface (or a sphere) must contain at least one country with five or fewer neighbors. His idea was then to suppose that there does exist a map which requires five colors. If this is so, then there must also be a map of this kind with the fewest possible number of countries contained in it. Considering this 'smallest' five-color map, he contemplated removing a border between two countries, thereby uniting them into a single country. The new map now has one fewer country than does the smallest five-color map. It must therefore be colorable using only four colors. It follows that putting the border back must require adding the fifth color. His idea was to examine all possible topological configurations that border replacement could require. This sounds a bit daunting, but you will see that a procedure can be readily developed to attack this problem in a methodical fashion.

Consider, for example, the possibility of having a country in the smallest five-color map that has three neighbors, and concentrate on this four-country configuration alone. When we remove a border we now have a three-country set which is part of a four-color map. If the four colors are red, green, blue and yellow then we can color these three countries red, green and blue, and still have the yellow left over to color the fourth country when the border is replaced. It follows that the smallest five-color map cannot possibly contain any countries with three neighbors. An exactly analogous argument also rules out countries with two or one neighbor from the smallest five-color map. So much for the trivial part. All that remains is to consider countries with four and five neighbors (since every map must contain a country with five or fewer neighbors). The case for four neighbors is a little more difficult and requires the examination of more than just the immediate neighbors. In particular, it becomes necessary to show that it is always possible to avoid coloring the four adjacent neighbors with four different colors. Kempe, however, was up to the task and was able to show that this was always possible. The real test came at the final hurdle—the country with five neighbors. For this case the number of different connectivities that have to be examined is immense. Nevertheless, after a painstaking search, Kempe believed that he had found an essential set of local patterns (essential in the sense that at least one of the patterns must occur in any map) and established that every single one of them could have the crucial border replaced without the necessity of adding a fifth color. The solution was announced and the mathematical world was duly impressed—firstly, with the complexity of such a simple-sounding problem, and secondly, with the technique and thorough computations of Kempe that completed the solution.

There the matter rested for a full decade until, horror of horrors, a single mistake was found in one of Kempe's computations concerning the country with five neighbors. One member of Kempe's essential set of local near-neighbor country patterns could not have the border replaced without the need for a fifth color; Kempe's demonstration that it could contained an error. This did not disprove the four-color conjecture since many other different essential sets of local configurations can be found. It merely established that it would be necessary to look for other essential sets of local patterns and test them as well. Unfortunately all the simplest essential sets (i.e. those involving the fewest countries) each appeared to include at least one example requiring the extra color. On the other hand it was becoming increasingly evident that the total number of different essential sets of near-neighbor configurations was both extremely large in number and extremely complicated in detail, and every one would need to be tested to prove the need for five colors (although any single 'lucky strike' could establish the sufficiency of four colors). It appeared that no single human (or even a sizeable group effort) would have the time or the fortitude to complete such a task in a lifetime.

So there the matter rested until the electronic computer came to the rescue. In fact the task at hand was so immense that not until 1972 was a computer with sufficient speed and memory available to seriously take up the challenge. This was the situation when Professors Kenneth Appel and Wolfgang Haken, of the University of Illinois, began their computerized assault. But still it required four more years of computer-program improvements and refinements before at last, in the summer of 1976, the triumph was announced. Over 1000 hours of computer time on three separate computers were needed to eventually pin it down. Over 1500 different essential-neighbor configurations had been tested (at least one of which had to occur in any given map) when finally, at last, one was discovered for which an extra border could be inserted in each and every component configuration without requiring an extra color. Thus, the smallest five-color map does not exist and four colors do indeed suffice, as the Post Office of Urbana (home of the University of Illinois) proudly stamped as part of its cancellation postmark on all mail leaving the city in the days following the announcement.

Clearly there are topological questions even in two dimensions that take us to the very cutting edge of mathematical research. And even though the question of whether a map (on a planar surface or a sphere) requires four colors or five is of little consequence to the world of science, the 120-year attack on this topological problem has led to the development of many new branches of mathematics with important consequences. One particularly important example is called 'graph theory', and concerns itself with the most efficient ways of connecting paths between points. How, for example, should we create telephone or airline routes that will serve the people in the most efficient manner? Such questions are of immense importance in today's transportation and communications industries.

Reverting to the historical development of the subject, the first book to contain the word 'topology' in its title was published in 1848. It was authored by the Göttingen Professor of Mathematics, J B Listing (1808–1882), and entitled *Vorstudien zur Topologie* or, in translation, *Introductory Studies on Topology*. Although there was not a great deal in it of true topological significance, aside from basic definitions and a number of properties of knotted curves, this was not so much an indictment of its author but rather a reflection of the fact that the subject was still in its infancy.

Riemann, who continued to prefer the term '*analysis situs*', was the first to introduce topological methods into the theory of functions. The basic task was to consider the behavior of analytic functions no longer restricted to a plane, but on other surfaces which may or may not be continuous, and to find out their topological properties using qualitative rather than quantitative methods—that is, without having to worry about their formal representation in the classical sense. Most of topology, in fact, deals with qualitative aspects, and in this respect it typifies a sharp break from earlier styles. A little later on in the nineteenth century, Poincaré was also to prove himself adept at handling functional or algebraic topology. Having been led to it by attempts at qualitative integrations of differential equations, he stated that practically every problem he touched led him to topology, and it was Poincaré who, in his book *Analysis situs* (published in 1895), spelled out the first systematic development of the subject. It was also he who was primarily responsible for extending the Eulerian topological studies of two- and three-dimensional polyhedra to higher-dimensional counterparts (or 'polytopes' as they are called) although the great impetus in the topological studies of invariants for these polytopes did not really come until the appearance of the electronic computer in the 1950s.

Today, topology is a highly developed, broad and fundamental branch of mathematics but (unfortunately for the typical scientist, whose mathematical education has not kept abreast of the ever more esoteric jargon and formalism of the subject) it has become progressively incomprehensible, bristling with less than friendly terminology and (at least to the uninitiated) appearing abstract in the extreme. To quote one of today's professional mathematicians, modern topological theory can be likened to 'a baby warthog, pretty to the few who love it, but of no interest to anyone else'. It is therefore not surprising that the topological mathematics that has thus far made an impact on the world of science harks back to the less abstract notions of the earlier years, and in particular to those concerning froths and knots.

One topological question that arises in several branches of physics concerns the nature and connectivity of the kinds of space-filling polyhedra that can be defined by an irregular set of points. But before focusing on this problem, which arises in many contexts from soap bubbles and foams to amorphous solids and glasses, let us first introduce the terminology via the much simpler question related to a regular, or ordered, array of points—such as atoms in a crystal.

Elemental solids (that is, those composed of a single atom-type only) tend to crystallize in a close-packed structure (see chapter 2). In fact, among the one hundred or so elements in the Periodic Table of Elements, more than half (including most metals and all five rare-gas crystals like helium, neon, etc) normally crystallize in such a structure. The simplest arrangement of this kind is the cubic close-packed (or face-centered-cubic) structure, a 'unit cell' of which is shown as the cubic F structure in figure 5 of chapter 2. If the sides of this cubic cell are (say) two units in length, then it is easy to see that any arbitrarily chosen 'center' atom at $(0, 0, 0)$, in Cartesian coordinates, is surrounded by 12 nearest neighbors at, respectively, $(1, \pm1, 0), (-1, \pm1, 0), (0, 1, \pm1) (0, -1, \pm1), (1, 0, \pm1)$ and $(-1, 0, \pm1)$. This number, 12, is called the coordination number of the lattice and is the simplest single numerical parameter that can, at least partially, characterize the structure. Somewhat more information concerning the geometrical nature of the cubic close-packed lattice can be found in a three-dimensional 'cell' defined to contain all the points that are closer to the arbitrary center atom than to any other (see chapter 2). This cell can easily be constructed by drawing the planes that perpendicularly bisect the lines connecting the center atom to all of its nearest neighbors. This is known to scientists as a Wigner–Seitz cell (after two prominent twentieth century solid state physicists) but is better known to mathematicians as an example of a Voronoi polyhedron, named after a mathematician of an earlier generation.

Georgy Fedoseevich Voronoi (1868–1908) was born and educated in Russia, but spent much of his adult life at the University of Warsaw. Among his many mathematical interests was one dealing with the determination of all the possible ways of filling a Euclidean space of n dimensions with identically shaped polyhedra. For any periodic crystal structure with a single atom-type, the Voronoi polyhedron is unique (that is, consists of a lone geometric object). Some examples are shown in figure 6 of chapter 2 where, with a little imagination, you will be able to spot the one appropriate for the face-centered cubic lattice. Voronoi himself was interested in spaces of arbitrary dimension, but only with packings involving identical polyhedra. These may all be related to periodic 'crystal' structures in the various dimensions. Difficult though this problem is, it pales when compared with the corresponding problem for non-crystalline structures. And it is in

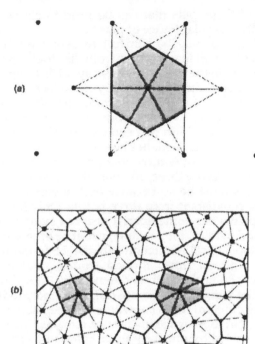

(a)

(b)

Figure 52

this context (albeit only in two and three dimensions) that the problem becomes a relevant one for the scientist in the study of random packings such as are found, for example, in metallic glasses.

Let us first examine the problem of defining general Voronoi polyhedra in two dimensions. In figures 52(*a*) and 52(*b*), for example, we contrast the situation for a close-packed regular triangular lattice and for an irregular, but still quite densely packed, array of sites. For the periodic lattice (whose points correspond to the centers of pennies packed in their densest possible configuration on a table) the Voronoi cell is a regular hexagon (figure 52(*a*)). In passing from this array to an irregular one, such as that shown in figure 52(*b*), we find that the unique and highly symmetric Voronoi cell associated with the former (and which fills the plane by repeating itself over and over as a honeycomb) is replaced by a statistical distribution of irregular shaped cells no two of which are, in general, identical. It is the distribution of these cells that can be used to characterize the topology of the irregular lattice of points.

At first glance it may be difficult to spot anything that is topologically invariant in a 'Voronoi froth' like that of figure 52(*b*); but it is there nonetheless and you may wish to spend a few moments looking at the figure and trying to locate it. Although each Voronoi cell is of a different geometrical shape, and some have four sides, some five, and some six or more, there is an invariant embedded within this complexity. If you look at each and every Voronoi cell, no matter how many sides and vertices it has, you will notice that, without exception, every cell corner (or vertex) is surrounded by three, and only three, cells. And it is easy to see why this must be so because each vertex is, by its very construction, equidistant from three points. Only if the original network of points has some special order or symmetry (as in a checkers board pattern, for example) can it be otherwise.

A similar topological restriction can also be found in a three-dimensional Voronoi froth, formed from a random (that is, unrelated) set of points distributed densely in three dimensions. Once again the individual Voronoi cells (now polyhedra) are all geometrically different, with anywhere between about 12 and 17 faces per cell, but topological invariance remains since now each edge is surrounded by precisely three cells, and each vertex by precisely four cells. The first condition follows because the locus (or path) of positions that are equidistant from three unrelated points is a line, and the second from the fact that the locus of positions equidistant from four unrelated points is itself a point. Although the situation is difficult to draw on a two-dimensional sketch, anyone doubting this fact can check it out at his or her leisure the very next time they play with soap bubbles in the bath. It remains true for any set of generating points that have no special set of relationships between them.

Thus, although geometrically each of the polyhedra in the froth is different in detail, we know something quite precise about them. Since three cells meet at each edge, it is evident that three faces must also meet at an edge. Similarly, if four cells meet at each vertex, then four edges must meet at each corner. It follows that any single Voronoi cell dissected from the froth must have two faces meeting at each edge and three edges meeting at each vertex. Let us now suppose that each face has p edges (and therefore also p vertices). With a total number V, F and E of vertices, faces and edges, respectively, it now follows that $3V$ and $2E$ must each count the total number of faces p times; or algebraically

$$3V = 2E = pF.$$

But we also know another relationship between V, F and E from the Euler topological result for an N-cell polyhedral froth, namely

$$V + F - E = 1 + N$$

where we can replace $1 + N$ by N on the right-hand side since N is an extremely large number. Solving these last two equations in a simultaneous fashion leads now to the relationship $F = 6N/(6 - p)$ or, for the average Voronoi cell making up the froth,

$$f = 12/(6 - p).$$

In fixing the definition of p above we have, for the moment, restricted our attention to cells with the same number of edges per face (namely p). Putting $p = 5$ (pentagonal faces) into the above equation gives $f = 12$, a condition that is satisfied by the regular dodecahedron (see figure 1 of chapter 2). Unfortunately, such an object doesn't fill space by repetition and therefore cannot be used to make a froth. In fact, 12-faced Voronoi polyhedra occur only rarely in three-dimensional froths.

In an infinite froth, the equation $f = 12/(6 - p)$ should be interpreted statistically with p standing for the *average* number of edges per face, and f the *average* number of faces per polyhedral cell. Expressed in another way we could write this as

$$\sum_p (6 - p)F_p = 6N$$

where F_p is the number of faces with p sides in the entire froth of N cells. Note that the equation sums over all faces and not all polyhedra; therefore a 6 instead of a 12 appears on the right-hand side since each face is shared between two polyhedra. This sum is a topological constraint for any three-dimensional froth of Voronoi cells. Although it is not a particularly strong constraint in three dimensions, the analogous constraint in two dimensions is very strong indeed, requiring the average number of edges per polygonal cell to be *exactly* six. Thus, the structure of an arbitrarily disordered network is actually significantly limited by topological requirements, and this result is of great importance in any attempts to model non-periodic structures in physics.

In chapter 2, for example, we described computer efforts to model metallic glasses in solid state physics. If we get a computer (using an appropriate packing algorithm) to assemble a dense random packing of equal-sized spheres, the characteristics of the

Voronoi polyhedra, corresponding to the 'froth' obtained from the sphere centers as the generating 'lattice', can be examined numerically with comparative ease. In the 1970s such computer models were indeed examined in detail in an effort to understand the structure of metallic glasses. Typically it is found that the number of edges per face runs all the way from $p = 3$ (triangles) to $p = 8$ (octagons), and the number of faces per polyhedral cell runs from about 12 to 17. The most common face is a pentagonal one (with about a 40% probability) followed by a hexagonal one (~30%) and a quadrilateral (~20%). For the number of faces per cell the most common is $f = 14$ (with about a 35% probability), followed by $f = 15$ (~25%) and $f = 13$ (~20%). Compared with the unique Voronoi cell for the face-centered-cubic lattice (with $p = 4$ and $f = 12$; see the top-right polyhedron in figure 6 of chapter 2) we see that the random froth tends to contain cells with both a larger number of faces and a larger number of sides per face than the unique Voronoi cell of the ordered lattice. It is found, in particular, that the average number of sides per face is about $p = 5.12$ while the average number of faces per cell is about $f = 13.6$. Thus, if a regular polygon were to exist that could exactly fill three-dimensional space in a randomly packed manner then it would have to have a non-integer number of sides. In other words, no such creature exists.

In many respects, the physics of glass structure became an extremely active area of physics in the 1970s, although a few pioneers had given some thought to the matter as early as the 1930s. As discussed in chapter 2, it happens that very few glasses have atomic structures that approximate a dense packing of spheres. However, all do possess a structure which is disordered in the sense of never repeating itself. Most glasses contain atoms that bond to each other via electron sharing. This so-called 'covalent' bonding fixes the nearest-neighbor configurations rather precisely as regards the number of nearest neighbors and their relative orientation. In fused SiO_2, for example, which we touched upon in chapter 2, each silicon (Si) atom is constrained by the requirements of quantum chemistry to have precisely four 'bonds' in a tetrahedral orientation. Each Si therefore has exactly four nearest-neighbor oxygen atoms positioned at the corners of a regular tetrahedron. Each oxygen (O) is also chemically constrained to bond to precisely two Si atoms. However, the energies involved in the bonding at the oxygen sites are not sufficient to fix the Si–O–Si bond angle precisely, so that this angle can therefore vary somewhat from oxygen to oxygen, thereby disrupting exact three-dimensional periodicity.

In cases like fused silica, with more than one species of atom and angularly constrained bonds, it is easier to focus on the network of silicon–oxygen bonds than to revert to the concept of a Voronoi froth. The topology of the resulting structure, a two-dimensional representation of which is shown in figure 8 of chapter 2, is best characterized by its ring statistics; that is, the frequency of occurrence of alternating silicon and oxygen bonded loops containing a particular number of atoms. Starting at a given atom, we trace a closed path of bonds, visiting each atom only once along the way and returning to the starting atom. In fused silica, for example, there are known to be a substantial number of eight-membered rings (each containing four silicon and four oxygen atoms). The exact ring statistics for real fused silica, as opposed to those for a computer-assembled model, are not known since no experiment has yet been carried out that is capable of discerning all these details with precision. Nevertheless, these statistics (whatever they are) do define the topological properties of the structure. That is to say, no matter how much we continuously bend, stretch or twist the amorphous lattice, these numbers will remain unchanged.

Regardless of whether we choose to describe disordered arrangements of atoms in terms of Voronoi froths or ring statistics, it is the topological measures of these descriptions that pin down the 'souls' of the structures. No two macroscopic samples of an amorphous metal or of fused silica will ever have even approximately comparable structures in the Euclidean or metric sense. On the other hand, their topologies, as measured by their short- and medium-range connectivities, will be the same (or at least extremely similar). And although even topologies may vary slightly from sample to sample (due to differences in the precise method of preparation, for example), it is evident that the entire field of amorphous material structure rests heavily upon the mathematical concepts of topology.

The connection, in fact, goes even deeper than this, since not all glasses are of either the close-packed or random network types. A third important class, thus far neglected in our discussion, is the class of polymer glasses. Polymers are long chains of organic molecules dominated by hydrogen and carbon atomic linkages. Being long they find it extremely easy to get tangled up, one with another, or even knotted. This being so, they are excellent candidates for forming non-crystalline solids (polystyrene being one example). Their structure is therefore intimately connected with the concept of 'entanglement'. In fact, perhaps the best-known simple model for such a polymer glass is known as the 'random coil' model—and random coils always imply the possibility of knots.

Knottedness is primarily a topological property of 'strings' in three-dimensional space. It cannot exist in two dimensions because there just isn't enough 'room' in two-space for a curve to become knotted, while in four dimensions there is too much room and nearly every reasonable embedded curve (that is, one which, at first sight, appears to be knotted) can be unraveled by continuous deformation. We all know what a knot is, of course, or at least think we do (especially if we were ever in the Scouts). But for mathematicians a knot is something just a little different from what we picture—more like an electrical extension cord that has been tangled up and then plugged back into itself. Thus, unlike the normal Scout variety, it always forms a continuous loop, and the manner in which it winds through space and crosses over and under itself before joining up is what makes all the difference to mathematicians. The study of knots is a relatively new area of mathematics with most serious work dating from the early decades of the twentieth century, although some less-formal efforts (such as Kelvin's attempt to deduce the Periodic Table of Elements by assuming atoms to be knotted vortices in the 'ether') can be found in the earlier literature. After long being something of a mathematical backwater, and certainly divorced from the world of applied science, the theory of knots has very recently become an extremely chic topic in physics with ramifications stretching beyond the realm of polymers even to the unlikely domain of statistical mechanics.

If one is to make any progress in the theory of knots, the first essential is to be able to tell whether two knots are topologically the same or different. Obviously, if one can, by bending, twisting or stretching, turn one knot into another then they are equivalent—but such a task quickly becomes a daunting one as the degree of complexity of the knot increases. Actually, mathematicians usually classify knots not by examining their three-dimensional form but rather by contemplating their two-dimensional shadows. In this projection, tiny breaks in the lines, signifying underpasses and overpasses, enable mathematicians to keep track of the knot's spatial placement (see figure 53 for some examples).

A loop without any crossings at all is topologically a circle and is sometimes referred to as an 'unknot'. Nevertheless, as you will quickly learn if you scramble up a long unknotted loop of string and 'flatten' it out onto a table, planar projections with many crossings can also, in reality, be unknots. The simplest true knot

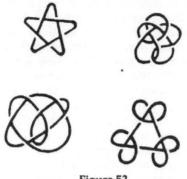

Figure 53

has three crossings and (always ignoring mirror images) there is only a single example, called the 'trefoil'. It is shown in figure 54, together with the lone knot with four crossings and the two knots with five crossings. The number of crossings in the simplest projection of a knot therefore seems to be a good starting point for seeking a knot classification. Unfortunately, as the number of crossings increases, the number of different knots with the same crossing number 'explodes'; there are, in fact, no less than 12 965 different simple (i.e. using only one length of string) knots with 13 or fewer crossings. So complex is the problem of knot formation that mathematicians have scarcely gone beyond the complete classification of 13 crossing knots (of which tabulation was completed in 1985) at the time of writing.

Figure 54

In addition to the classification of simple knots, we are also confronted with the question of how to reduce complicated-looking knot shadow projections to their simplest (i.e. with fewest crossings) form, and of confronting the entirely separate field of knots involving more than one loop (called 'links'). Because the possibility always remains that some clever trick of twisting will deform two seemingly different knots or links into each other, it is highly desirable to find some property that is unambiguously capable of distinguishing them—that is, some numerical or algebraic expression that can be assigned to each knot and which remains the same no matter how much the knots are continuously deformed. Such a property is called a 'knot invariant' and a primary focus of attention in knot theory throughout this century has been the effort to locate an invariant that will be a sufficient condition to uniquely identify any knot. This ultimate invariant still eludes the experts, but many less powerful invariants have been found. The most used is called the Alexander polynomial, discovered by the American mathematician James W Alexander in 1928, but many others have been located over the years, the most recent (discovered in the mid-1980s) being rather amazingly derived from the statistical mechanical theory of the up–down spin magnet that was described in the last chapter. This intimate connection between statistical mechanics and knots (noted first by Vaughan Jones, of the University of California, Berkeley) is still something of a mystery, but is particularly exciting since it promises to forge a close connection between two seemingly quite disparate fields.

The primary interest of physicists and chemists in knot theory is in terms of its rapidly developing connection with the science of polymer chemistry. Both chemists and biochemists have been able to synthesize polymeric rings which are knotted in complex ways. A great deal of effort is now being expended to study these knotted systems since both their dynamic and thermodynamic properties seem to be sensitive functions of their topological state. One of the biggest practical growth areas is molecular biology. Knotted forms of a number of large natural polymers have been known now for close to two decades and chemists are now able to tailor-make simple knots in synthesized (that is, laboratory-constructed) molecules. The most famous of these structures is the DNA molecule which carries within it the genetic code governing life itself. In the vital life processes of replication and recombination of DNA molecules (which are performed by entities known as enzymes) the DNA gets cut into sections and reglued in sequences of steps in which one structure is gradually turned into another. It is now known that these enzyme activities on DNA are the very stuff of which modern knot theory is made. When reacted with unknotted (i.e. topologically circular) DNA each enzyme produces a characteristic family of knots that can be experimentally observed by high-resolution electron microscopy.

Theoretically, the simplest such system is the single ring polymer (like DNA) and it is now possible to model such polymers through unbiased computer simulation to see how the degree of entanglement varies with the length of the polymer. As the chain length increases the probability of having an unknotted arrangement appears to decrease exponentially with chain length. In addition, the kinds of knots that appear, and their frequency of appearance, have been analysed in some detail. Ring lengths of up to several thousand polymer bonds have been analysed and knots with up to ten crossing points identified. It now seems only a matter of time before linked polymers and polymer glasses can also be analysed to clarify the role played by knots in determining the properties of these even more complicated materials as well. And although entanglement does not necessarily imply knots, it does now seem extremely likely that at least a fraction of the polymer strings in, say, plastics are knotted and that the marriage of polymer theory and the topological theory of knots will be a long and fruitful one.

One other area in which topology is making surprisingly valuable inroads in science is in the prediction of the properties of complex chemical substances before they have ever been synthesized. More and more it is becoming apparent that it is the manner in which the component atoms are linked, rather than their three-dimensional shapes, that play the essential role in this context. The topological analysis of a molecule begins with a drawing in which the molecule's atoms are depicted as points (vertices) and the connecting chemical bonds as lines (edges). The length of the lines and the angles between them do not matter. Drawings of this kind are known as chemical graphs. Graphs of this structural nature were first studied in the nineteenth century by the Cambridge mathematician Arthur Cayley. His work spawned the now vibrant mathematical field called graph theory, and chemical graphs are the basic tool used for applying graph theory to chemistry.

Once a chemical graph has been drawn it is important to search for topological invariants in the structure. The first such topological index that met with significant success in predicting chemical properties was put forward in 1947 and is referred to as the

'Wiener' index, in honor of its discoverer. If the 'distance' between any two vertices is counted merely as the smallest number of edges joining them, then the Wiener index of a molecule is the sum of all such 'distances' between every pair of atoms in the molecule. The index obviously increases with the size of the molecule, but it also contains within it a measure of the connectivity or branching structure present. This Wiener index correlates surprisingly well with such properties as boiling point, surface tension, viscosity, and even optical properties. In more recent years, other more intricately devised numerical invariants have been derived which can perform even better in this context. They have been able to model an incredibly wide range of chemical, physical and even biological phenomena with surprising accuracy. What this is telling us is that it is the pattern of interconnections among atoms, more than anything else, that appears to dominate the chemistry of complex molecular structures. In short, it is becoming more and more apparent that topology is not only the soul of geometry but also, to an ever increasing degree, the soul of polymer chemistry too.

11

From Parabolas to Fractons

One of the greatest advances ever made in geometry was achieved long ago with the discovery of its intimate connection with algebra via what we would now refer to as analytic geometry. Although, as spelled out in chapter 3, there is no unanimity of opinion as to who first appreciated this connection, the idea of fixing the position of a point on a plane by means of choosing a pair of numerical 'coordinates' (a, b), and of interpreting algebraic equations relating a to b in a geometrical fashion, certainly goes back at least to the Ancient Greeks of the third century BC. So comfortable have we become with the notion of analytic geometry that most of the formal proofs set out by the early Euclidean geometers now appear to be almost painfully tortuous and pedantic. Today, symbolic algebra and trigonometry have almost completely replaced the original pure geometric formalism. Thus, for example, Euclid's statement that 'if a straight line be cut at random, the square on the whole is equal to the squares on the segments and twice the rectangle contained by the segments' would now be thought of as an extremely cumbersome manner of expressing the simple algebraic identity

$$(a + b)^2 = a^2 + 2ab + b^2.$$

The essence of analytic geometry, as applied to the plane, is the association of the position of any point in the plane with an ordered pair of numbers which are measured with respect to an arbitrarily chosen point $(0, 0)$ that is selected to be the 'origin'. In the more conventional notation (x, y), x then measures (to use a geographical analogy) the distance to the east (more formally known as distance along the 'abscissa') and y measures the distance north (or along the 'ordinate') that must be traveled to get from the origin $(0, 0)$ to the point (x, y). In this scheme, any algebraic relationship between x and y now defines a curve in the plane. Thus, for example,

$$x^2 + y^2 = 1$$

defines the circle of radius 1 centered at the origin. To verify this it is only necessary to plot on a piece of graph paper all the separate pairs of real numbers x and y that satisfy the equation. More generally, if we can conceive of any relationship between x and y such that when x is given y may be calculated, then y is said to be a 'function' of x, and the resulting relationship, which we can write as $y = f(x)$, can be 'translated' geometrically.

The simplest such relationships result in geometric forms that are continuous curves, and studies of analytic geometry usually start with such standards as a straight line

$$y = mx + c$$

with m and c as constants, and a parabola

$$y^2 = ax$$

with 'a' as a constant. The notion that one could use this relationship between geometry and algebra to actually prove geometric theorems by algebra came to the forefront in the early-seventeenth century, and is usually credited to Descartes and Fermat. The idea was simply to transform any proposed geometric theorem into its algebraic representation and to establish the validity or falsity of the statement by algebra alone. Conversely, this ability to switch back and forth between geometry and algebra also means that new algebraic findings (derived solely within the confines of mathematical analysis) may now very well translate into new and unsuspected geometrical results.

In the early days of analysis it was thought that any function that was defined by a simple mathematical formula and which did not become infinite anywhere (as does $y = 1/x$, for example, at $x = 0$), would be continuous and smooth. By continuous, as Euler later suggested, one meant a curve which could be traced in its entirety without lifting the pen from the paper. It is easy to establish the demise of this part of the above speculation by considering the geometrical representation of the function

$$y = x + \sqrt{(x - 1)(x - 2)}$$

which abruptly 'disappears' between $x = 1$ and $x = 2$ (see figure

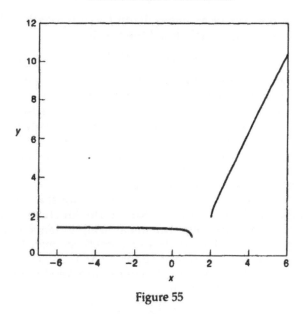

Figure 55

55). The smoothness part (when the curve existed) seemed to be on rather firmer ground and it was just such a concept that led Newton, in the latter half of the seventeenth century, to develop the notion of slope at a point, and differentiation (see also chapter 12). For example, he was able to consider the slope of a curve like $y = x^m$ at the point x by envisaging arbitrarily small increments dx and dy in x and y, respectively, and calculating the ratio dy/dx as these increments were allowed to become vanishingly small. Writing

$$y + dy = (x + dx)^m$$

he was able to expand the right-hand side, using a theorem (the so-called 'binomial' theorem) which he himself had discovered in 1665, as follows:

$$y + dy = x^m + mx^{m-1}\, dx + \text{(terms in } dx^2, dx^3, dx^4, \ldots)$$

where we shall have no need for the explicit representations of the higher-order terms. Subtracting $y = x^m$ and then dividing both sides by dx now leads to a slope between x and $x + dx$ of

$$dy/dx = mx^{m-1} + \text{(terms in } dx, dx^2, dx^3, \ldots).$$

Although Newton did not use this exact notation, he did

introduce the idea of allowing dy and dx to (together) become arbitrarily small, in which 'limit' the result

$$dy/dx = mx^{m-1}$$

represented the slope of the curve $y = x^m$ *at* the point (x, y). Using examples of this kind, Newton found that the analysis was just as self-consistent for values of m which were not positive integers (and for which the binomial series was of infinite length) as it was for the more familiar expansions with $m = 2, 3, 4, 5, \ldots$, and fractional and negative powers m no longer bothered him. Within this picture a curve was 'smooth' so long as the function representing the slope was continuous.

Using this method, simple rules for finding the slope of (or 'differentiating' to use the more formal term) functions were soon set up and are now thoroughly familiar to all students of calculus. However, this very familiarity can sometimes lead to disaster since it nowadays replaces thought all too frequently. As an example, I like the story of the teacher who asked his students to find the slope of the function

$$y = \log [\log (\sin (x))]$$

at the point $x = \pi/4$. If you blindly follow the standard rules of differentiation you obtain the answer

$$dy/dx = \cot (x)/[\log (\sin (x))]$$

which, at $x = \pi/4$, takes the value $1/(\log(1/\sqrt{2}))$ or about -6.64. Having found the slope via the well documented rules, the teacher then suggested that it might be interesting to sketch the complete function $y = \log [\log (\sin (x))]$ to see whether this value of slope looked 'reasonable'. Do you wish to give it a try? It doesn't take long. Since $\sin(x)$ is always less than or equal to one, it follows that $\log (\sin (x))$ must always be less than or equal to zero. Now, since the logarithm of zero or a negative number does not exist, it follows that $\log [\log (\sin (x))]$ also does not exist for any value of x. The whole function is a fraud—and yet it seems to have a perfectly well defined slope at most values for x, and at $x = \pi/4$ in particular. We have here something like Lewis Carroll's Cheshire cat—a function which can vanish and leave its slope behind it. I leave those of you who consider yourselves to be experts in the field of calculus, and quite complacent about simple differentiation, to wrestle with this conundrum. As a clue, the statement that the logarithm of a negative number doesn't exist is true only if we restrict ourselves to real numbers. That is, a full understanding requires a knowledge of functions of a complex variable and, since complex numbers did not put in an appearance in mathematics until the turn of the nineteenth century, such was not available to Newton. Whatever the answer, it is clear that not *all* analytic functions $y = f(x)$ of a real variable x can be translated into a geometric representation. It seems natural then to ask whether the reverse might not also be true. Are there, perhaps, geometric shapes that have no analytic representation?

The first such shapes to cause mathematical consternation were the ones like those shown in figures 14 and 15 (chapter 4). These are curves with sharp corners, and for years mathematicians expressed a great dislike for them, even debating as to whether they should be classified as 'functions' at all. And then, very early in the nineteenth century, Fourier (whose story is set out in more detail in chapter 4), while studying an analytic approach to the theory of heat flow in solids, found a method for expressing disjointed 'curves' of this kind as an infinite series of sinusoidal oscillations. Although most of Fourier's esteemed French mathematical colleagues (including Lagrange, Legendre, Laplace and Poisson) found his claims difficult to accept in full, it did indeed seem that an infinite series of homely trigonometric functions might well be sufficient to give birth to many of these weird geometric creatures with sharp corners. The primary difficulty, as far as rigor was concerned, involved the notion of 'convergence' or 'limit' in the context of infinite series.

The question of convergence arises whenever an infinite sequence of numbers is to be added or multiplied (or subjected to some other mathematical operation). For example, most of you will be able to convince yourselves that the series

$$S = 1 + \tfrac{1}{2} + \tfrac{1}{4} + \tfrac{1}{8} + \ldots$$

'converges' to the limiting value of 2. But can you set down a convincing proof? And if you can, what about the series

$$S = 1 + \tfrac{1}{2} + \tfrac{1}{3} + \tfrac{1}{4} + \ldots$$

or the seemingly paradoxical

$$S = 1 - 1 + 1 - 1 + 1 - 1 + \ldots.$$

Is the latter, for example, equal to

$$S = (1 - 1) + (1 - 1) + (1 - 1) + \ldots = 0 + 0 + 0 + \ldots = 0$$

or rather to

$$S = 1 - (1 - 1) - (1 - 1) - (1 - 1) - \ldots = 1 - 0 - 0 - 0 - \ldots = 1?$$

Or should it even be calculated according to

$$1 - S = 1 - (1 - 1 + 1 - 1 + 1 - 1 + 1 - 1 + \ldots)$$
$$= 1 - 1 + 1 - 1 + 1 - 1 + 1 - 1 + 1 - \ldots = S$$

in which case, $1 - S = S$ and therefore $S = \frac{1}{2}$?

Clearly a blind manipulation of terms within infinite series can lead to mathematical inconsistencies of a most embarrassing kind and, therefore, series of this kind must be treated with extreme caution. It was problems of just this nature that led many of Fourier's colleagues to have reservations about his all-embracing claim—namely that any geometric function whatsoever, whether smooth or jagged, finite or infinite, can be expressed as a (possibly infinite) series of sines and cosines. And, as detailed in chapter 4, these reservations were not totally unfounded although it took the better part of another one hundred years to finally settle the question. It turns out that there are, indeed, limitations imposed by the convergence requirements for, although 'Fourier analysis' is applicable for *most* functions (even ones with pointy corners), one can conceive of some rather pathological curves for which no Fourier series exists. Paramount among these Fourier renegade curves are the kind that contain an infinite number of discontinuities in a finite interval. But who in his right mind would ever have the need to concoct a geometric curve like that?

Certainly, for most of the nineteenth century, functions of this nature were given very little serious attention by the practitioners of mathematics, and none whatsoever by scientists. If no analytic expression existed that could represent them then they must, in some sense, be pathological and certainly of no relevance to an understanding of the real world. The general sentiment seemed to be that if Euclidean geometers could not draw them with a rule and compasses, and analytical geometry was equally confounded by their mathematical expression, then they could surely be pre-empted from serious mathematical discussion as well.

The first crack in this attitude appeared in 1890 when the Italian mathematician Giuseppe Peano (1858–1932) presented one of the more disquieting discoveries of the time. Specifically, he demonstrated the manner in which a simple mathematical procedure could lead to a curve which outraged common sense by completely filling a finite area of space. In fact, many such curves could easily be constructed (at least in principle) and all examples

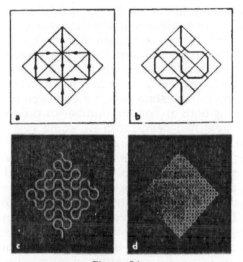

Figure 56

of this area-filling type of curve are now referred to collectively as 'Peano curves' in honor of their discoverer. Let us set out an example to see how this sort of thing can happen.

Take a look at figure 56. The procedure for constructing this particular Peano curve begins with a square and the dividing of each of its sides into three equal segments. In this manner one can imagine the original square divided into nine smaller squares as set out in figure 56(*a*). The Peano curve construction then starts by drawing a straight-line diagonal across the large square from the bottom to the top and then replacing it by a continuous line through the diagonals of all nine of the smaller squares, following the pattern of arrows in the figure. In order to help in seeing the shape of the resulting curve we now chop off the corners of each right-angled twist, as shown in figure 56(*b*). Neglecting this corner-chopping in our formal development, we note that each of the nine small squares now has a straight-line diagonal across it and therefore looks exactly like the original square when we started. We can therefore split up each small square into nine even smaller ones and repeat the original procedure in each. The result is a curve that now has $9^2 = 81$ (corner-chopped) line segments and it is drawn in figure 56(*c*). But now, of course, the building procedure can be repeated yet again to extend the curve to one with $9^3 = 729$ segments (figure 56(*d*)) and so on, at least in principle, *ad infinitum*.

Looking at the sequence of developmental curves in figure 56 it is, I think, fairly evident visually that when continued to the infinite limit of line fragmentation this curve will completely fill the area of the original square. But a line, according to Euclid's definition, is a breadthless length. How can a breadthless length fill up an area? Are there not infinitely more points in an area than on a line? This last question, which involves counting the infinite (since there is both an infinite number of points on a line and within an area) had been taken up some years earlier by the German mathematician Georg Cantor (1845–1918). His work, which was equally controversial at the time, provided a method for distinguishing between different sorts of infinite quantity. Using this method he was able to establish that there are the same infinite (or, to use the formal term, transfinite) number of points in an area, (and, yes, even a volume) as there are on a line, but that this number is infinitely larger than the total number of integers 1, 2, 3, 4, . . ., which, of course, is also infinite. Cantor established that there are many different kinds of infinity (in fact, an infinitude of them), but that the number of points on a line, area, or volume corresponds to the very same one.

Peano's curve did not therefore violate the laws of the infinite, even if it did present something of an outrage to common sense. The rational mind can only make judgements based on finite quantities since it only has experience of the finite. Any finite density of points will always place less points on a line than inside an area containing the line, and always put less points on a short line than on a long one. But transfinite numbers behave quite differently and for them two infinite sets of any kind are equal if, and only if, the members of one can be placed in a one-to-one correspondence with the members of another. Thus, for example, there are exactly as many total integers as even integers. It pays to ignore 'common sense' in such circumstances.

Peano's curve, and the controversy concerning it, was followed in 1904 by an equally pathological, but far more esthetically appealing, curve constructed by Helge von Koch (1870–1924) of Stockholm, Sweden, whose major contributions to mathematics were associated with infinite systems and limits. Usually referred to as the 'Koch snowflake', it can be constructed (see figure 57) by beginning with an equilateral triangle and marking off each side into three equal parts. Each center section is then removed and replaced by two sides of another smaller equilateral triangle for which the third side is the now-removed center span (figure 57). The resulting shape is the 'Star of David'. Once again, as for the Peano curve, the generating procedure can be repeated upon

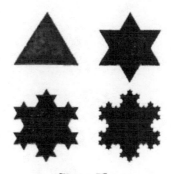

Figure 57

each of the smaller line segments, and then repeated again, and again *ad infinitum*. Figure 57 shows only the next two iterations following the Star of David after which the resulting 'snowflake' becomes more and more difficult to draw as the 'spikes' get closer and closer together. Carried on (in principle) to infinity, there is ultimately a spike at every point of the curve. The snowflake has no smooth regions at all; no tangent can ever be drawn anywhere on it and no slope ever calculated. And yet, unlike the Peano curve, it doesn't fill up any regions of two-dimensional space. In fact, it contains a perfectly well defined area inside it although the distance between any two points along the curve can be shown to be infinite. Thus, it certainly doesn't behave like any normal one-dimensional curve; and yet, equally clearly, it isn't two-dimensional either. What then is it? Could it be something in-between dimensions—and does it really matter to anyone out-side the field of pathological geometry?

The idea that objects like this snowflake curve really could be construed as existing in the realm between one and two dimensions was first put forth by the German mathematician Felix Hausdorff (1868–1942), who was also a pioneer in the algebraic development of topology. Topology, as you will recall from the previous chapter, deals with properties of geometric figures that are not altered under continuous deformations of space. As an offshoot of this line of study Hausdorff, in 1919, first introduced into mathematics the property of 'scaling'. Scaling is the process of sequentially measuring the same object with different units of measure and, using it, Hausdorff was able to suggest a manner in which it was possible to generalize the notion of dimension in a consistent fashion, thereby putting these 'monster curves' into a respectable mathematical classification.

Felix Hausdorff, 1868–1942. Courtesy of Mathematisches Forschungsinstitut, Oberwolfach.

The basic technique is easily demonstrated. Consider, for example, a square with perimeter P and area A. If we measure the perimeter using a measuring rod of length L, then we shall deem it to have a length

$$P(L) = P/L$$

just as a 'yard stick' measured with a 'one foot' rule would be given the 'length' three. Measuring the same perimeter with a new measuring rod of length L/N now results in the new measure

$$P(L/N) = NP/L = NP(L).$$

We can do the same thing for area measurement as well. First, using an 'L by L' square measuring unit, we find an area

$$A(L) = A/L^2.$$

Measuring again, this time with an 'L/N by L/N' measuring unit, we obtain

$$A(L/N) = N^2A/L^2 = N^2A(L).$$

Clearly, for volumes V of three-dimensional objects an analogous relationship

$$V(L/N) = N^3 V(L)$$

can be established.

Looking at these three scaling relationships involving, respectively, linear, area, and volume measure, it is evident that the conventional notion of dimension conforms with the exponent of the 'scaling factor' N in each case. And where does this leave the Koch snowflake? Adding a condition that one cannot measure any detail smaller than the unit of measure (scientists may understand this unit as the limit of experimental resolution), it is not difficult to see that for this curve (figure 57) each time we reduce our unit of measure by a factor of three the perimeter measure increases by a factor of four. In other words, for the Koch snowflake we can write the relationship

$$P(L/3) = 4P(L).$$

If we now rewrite this in the form

$$P(L/3) = 3^d P(L)$$

to conform with our notion of dimension d as defined above for one-, two-, and three-dimensional objects, then we deduce a dimension

$$d = \log (4)/\log (3) = 1.261\,859 \ldots$$

for the snowflake curve.

For the Peano curve of figure 56 we may follow a parallel line of reasoning to bolster our earlier visual conception that it is, in the iterative limit, a two-dimensional curve. We see from its method of construction that each time we reduce the unit of (linear) measurement by a factor of three, the perimeter measure increases by a factor of nine. It follows that

$$P(L/3) = 3^d P(L) = 9P(L)$$

from which we find a dimension $d = 2$ in conformity with our visual expectations.

The word now used to designate 'between-dimension' curves is *fractals*, with d being their Hausdorff (or fractal) dimension. The word 'fractal' was coined in the mid-1970s by the Polish-born American mathematician Benoit Mandelbrot. It has, as its root, the Latin word *fractus*, meaning irregular—and these fractal curves are certainly the epitome of irregularity, albeit irregularity

Benoit Mandelbrot, 1924–. Courtesy of B Mandelbrot.

of a very special kind. This special character concerns the fact that they are all 'self-similar' in the sense that 'zooming in' for a closer view does not smooth out the irregularities. Instead, these objects exhibit exactly the same degree of roughness no matter what the level of magnification. Nor are they rarities. In fact they are extremely easy to create. Pointing triangles inward rather than outward at each step of the Koch snowflake construction creates an equally mysterious object, but one could just as well add pieces of any shape to any starting form so long as a regular building procedure is established which can repeat itself exactly at each stage and be continued *ad infinitum*.

Fractals can also be formed in three-dimensional space as easily as on a plane or a surface. For example, dividing each face of a regular tetrahedron into four equal-sized equilateral triangles and erecting a tetrahedron on the middle triangle of each face also begins a step sequence which, when continued to infinity, creates an infinitely prickly higher-dimensional analog of the Koch snowflake. Its surface has infinite area but it encloses a

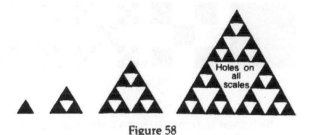

Figure 58

finite volume. And since its surface measure increases by a factor of six each time the 'unit' measuring triangle decreases its side length by a factor of two, we can write

$$A(L/2) = 2^d A(L) = 6A(L)$$

to deduce its fractal dimension as $d = \log(6)/\log(2) = 2.584\,96\ldots$.

Fractals, however, are not restricted just to infinitely prickly curves or surfaces. One of the joys of fractal geometry is the opportunity to create all manner of hideous or beautiful forms (depending upon how you see them). All that is needed is a basic starting shape and a rule which step-by-step makes the figure more irregular on ever smaller scales in an endless iteration (or repeating process). As an example of a fractal of a somewhat different nature from the above 'pricklies' we shall now introduce the Sierpinski gasket (named after its inventor, the Polish mathematician Waclaw Sierpinski (1882–1969)). Its rules of construction are apparent from the sequence of patterns shown in figure 58. At every step, three equilateral triangles are stacked together to form a larger equilateral triangle with a hole in the middle, the linear size increasing by a factor of two. The reader may imagine the sequence of steps shown in figure 58 as a result of observing the same object under ever-increasing magnification, so that ever more internal structure becomes visible. Each time the unit of measure (the side length of the smallest discernable triangle) is decreased by a factor of two, the total number of triangle sides increases by a factor of three. The fractal dimension is therefore $d = \log(3)/\log(2) = 1.584\,96\ldots$. Another fractal in the same vein, but this time based on squares, is shown in figure 59. At each step every square is replaced by five smaller squares as shown. Its dimension is $d = \log(5)/\log(3)$ or about 1.4650.

The methods for constructing objects of this kind are readily extended to three dimensions. Thus, for example, as a higher-dimensional analog of the Sierpinski gasket, four identical tetra-

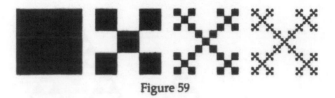

Figure 59

hedra can be joined together to form a larger porous tetrahedron at each step. The fractal dimension of this infinitely porous 'sponge' is then seen to be $d = \log (4)/\log (2)$, or exactly 2. Despite its very different nature this therefore has the same fractal dimensionality as the Peano curve of figure 56. The Peano curve is said to be both topologically and fractally of dimension two, while the sponge gasket is topologically three-dimensional. By this we mean that the Peano curve can be 'drawn' on a two-dimensional surface while the sponge gasket requires a space of three dimensions for its topological construction.

Fractals can also be defined with an Hausdorff dimension of less than one. An example, called the Cantor bar (after Georg Cantor of transfinite number fame), is shown in figure 60. Starting with a bar (or straight line if you prefer) we first divide it into three parts and remove the center section completely to obtain two shorter bars. The shorter bars are then each subjected to the same subdivision and the procedure continued indefinitely. In this fashion we finish up with a bar or line composed of an infinite number of infinitesimal fragments, each separated (at least to some degree) from its neighbors. The fractal dimension of this hazy object is easy to deduce by the now-familiar method and is $d = \log (2)/\log (3) = 0.630\,93 \ldots$.

Quite evidently these fractal objects are not difficult to invent. Nevertheless they remained nothing more than a mathematical oddity until well into the 1960s when Mandelbrot introduced a revolutionary notion. It was that, contrary to the beliefs of earlier generations of scientists who had always tried to understand

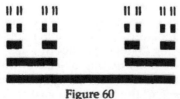

Figure 60

nature in terms of analytic functions and 'well behaved' geometries, the real world actually abounds with fractal objects. In his now famous words 'clouds are not spheres, mountains are not cones, coastlines are not circles, bark is not smooth, nor does lightning travel in a straight line'. In fact, each of these objects contains within it a fractal identity with a well defined Hausdorff dimension, even though the natural examples (called random fractals) are self-similar in a less precise fashion than their geometric cousins, and their self-similarity must eventually terminate short of '*ad infinitum*' (by virtue of the finite size of the molecules of which they are composed if nothing else). However, before progressing to the physical manifestations of fractal forms, let us allow ourselves one more peek at the magic of mathematical fractal patterns, this time as generated by a somewhat more subtle technique involving complex number iterations.

The complex plane, which is inhabited by complex numbers like $z = a + bi$, is really just an ordinary plane in which a and b measure respectively the x and y Cartesian coordinates of the point z, as discussed in chapter 5. Complex numbers, you will recall, add, subtract, multiply and divide, just like ordinary algebraic number forms so long as we replace i^2, wherever it arises, by minus one. Let $z = a + ib$ and $c = e + if$ be two complex numbers. Suppose that we square z (i.e. $z^2 = a^2 - b^2 + 2abi$) and add it to c to give the complex number $z' = (a^2 - b^2 + e) + (2ab + f)i$. The original point has now 'jumped' to another with an x-coordinate of $a^2 - b^2 + e$ and a y-coordinate of $2ab + f$. This mundane exercise springs to life when it is iterated; that is, computed repeatedly with the endpoint z' of one operation being used as the start z of the next. The resulting sequence of complex numbers does a strange jig on the complex plane and, as the point wanders, it is interesting to ask whether it will eventually rush off to infinity or dance forever in some confined region. In a more picturesque language we may ask 'Will the prisoner escape or not?' But how is one to select the initial values for z and c to start the iteration and look at the possibilities. One method is to make the initial z-value equal to zero and follow the excursions of the iterated numbers for different selected values of c. If this is done we find that for some c-values the 'prisoner' escapes while for others he doesn't. Coloring c white if the escape is successful and black if it is not now creates a pattern on the complex plane. The point of interest is the fact that this black 'prison' turns out to be a shape of essentially infinite complexity—see figure 61—with a boundary (now called the Mandelbrot set) of unbelievably intricate fractal form. The figure shown cannot really do it justice for,

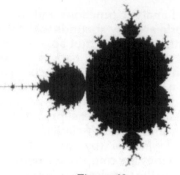

Figure 61

if you zoom in on it at ever higher levels of magnification (and you need a very fast computer to do this with ease), each new level reveals new and ever-surprising structures open to every imaginable interpretation in the mind's eye. The fractal dimension of this fascinating object is by no means obvious, although very recent work on the problem claims to prove that it is $d = 2$.

If the starting z is set equal to some non-zero value, other equally fanciful but related fractal patterns appear in which people have been known to identify sea horses, rabbits, sparks, whirlpools, snakes, bushes of brambles, and countless other shapes. However, under the opposite rule (when c is fixed and z plays the role of initial parameter) some of the resulting patterns can look rather similar to the Mandelbrot set while others are startlingly different. Their boundaries are called Julia sets, after the French mathematician Gaston Julia (1893–1978) who first studied these weird manifestations as early as the First World War years without, of course, any assistance from electronic computers or computer graphics! For some values of c the sets are connected (i.e. all in one piece like the Mandelbrot set) while for other values they are not. In fact, as a function of a linear progression of c across the complex plane, Julia sets can often be seen to shrink from connected Mandelbrot-like sets to fragile skeletons before 'exploding' into fractal dust. In all of this can be perceived associations with the 'strange attractors' of iterated chaos as discussed in chapter 7, and there are indeed close relations between Mandelbrot and Julia sets and the period-doubling route to chaos.

The closer we look at objects in nature, the more evident it becomes that most of them lack smoothness in a very complete sense, that is to say they seem to possess an irregularity no matter how small a measuring unit one adopts (short of atomic dimensions). Perhaps Mandelbrot's most famous early examples were coastlines. As a space traveler approaches, bays and headlands are first perceived. Then, on closer approach, these bays and headlands develop structures of their own. On the scale of a yardstick the details of rock formations need to be recognized, and on the centimeter scale the positions of pebbles have to be noted. And further beyond, if the sea could be stilled, even grains of sand would contribute to the picture. Measuring the length of a typical coastline P in this fashion as a function of the length of a measuring rod L/N (with N as the variable) generates the classic fractal form

$$P(L/N) = N^d P(L)$$

with d of the order 1.2 to 1.3.

In the case of coastlines, d is an experimentally deduced quantity and is therefore never known exactly. Unlike the Koch snowflake, the coastline is self-similar in only a statistical sense. The implication is that self-similarity persists upon magnification in a manner that involves some averaged property (namely the fractal dimension) over and above all the random changes of detailed form. Coastlines, and most naturally occurring fractals, are therefore referred to as 'random fractals' as opposed to the 'ordered fractals' of most mathematical fractal constructs.

Random fractal analogs of Cantor sets can be found in nature all about you — in the form of tree structures. The abstract structure of any tree is given by its branching properties, with larger branches splitting into smaller branches, smaller branches into twigs, twigs into 'twiglets' and so on. Although this behavior in real trees ceases at some small, but still macroscopic, length scale, it is easy to envisage the process continuing to the infinite limit for which we approach a 'fractal canopy' which distributes an infinite number of isolated points across a circular arc of finite size. This same fractal tree structure can be found in many other physical guises. Our lungs and blood vessels are organized in this same fashion; many forms of electric breakdown as voltage is increased across an electronic insulator also occur as fractal trees, and many crystals can be made to grow from seed in this fashion if the growth velocity is properly adjusted. Fractal tree structures are also created through what is known as 'viscous fingering' when one fluid, which is confined in a thin layer between two plates, penetrates into another fluid. Whether there is a common physical principle that is responsible for creating these fractal tree structures in so many different contexts is still a point of contention. After all, the physics of the growth (or dynamics) of fractals is a subject that is really only just beginning to emerge. Does the fractal tree perhaps provide an optimal transport method for distributing particles or fluids from a single point to a larger region? Questions of this kind are at the cutting edge of today's fractal research.

Fractals also play a central role in the physics of what is known as 'percolation'. The concept of percolation can be very simply illustrated by thinking of a chess board of infinite size for which the squares, instead of being colored black and white in the traditional alternating fashion, are each colored either black or white in a random fashion, with a probability p of being black. If p is small, say 1/10, then only one tenth of the squares will be black. The pattern which they create will then consist primarily of isolated black squares ('isolated' implying the sharing of edges with only white squares) together with a few small clusters of perhaps two or three adjacent black squares. However, as the probability p of being black is increased, the number and size of the black-square clusters also increase, with previously isolated squares and small clusters fusing together with others. If we focus upon the largest cluster of all, there eventually comes a time (i.e. probability value p) for which it becomes infinite. This very special value for p, which we symbolize by p_c, is then called the critical percolation concentration.

The percolation problem is not, of course, confined to chess boards nor even to problems in two dimensions, and p_c will take on different numerical values depending on the dimension and topology of the lattice involved (for interest, its value for the infinite chess-board configuration is close to 0.592). However, the important point in the present context is the fact that it can be proved that the *shape* of the infinite cluster at the percolation limit is always a fractal. And in scientific contexts it is a most important fractal since the phenomenon of percolation occurs in many different fields of research. As a first example, let us consider the case of non-magnetic atoms being gradually replaced by magnetic ones in a crystal. Since, in the simplest model of their description, magnetic atoms can interact strongly only when they are nearest neighbors, and behave like tiny atomic bar magnets (lining up north to north and south to south when they are neighbors, but remaining orientationally uncorrelated when they are more distant), we only obtain 'bulk' magnetism when the infinite cluster, or critical percolation, concentration limit of magnetic atoms is reached. It follows that, near $p = p_c$, all magnetic properties in these kinds of system are governed by the properties of fractals.

Other examples of percolation from the field of physics are the flow of liquids in a porous medium (flow will start only when the percolation limit is reached), the electric breakdown of a crystal when subjected to larger and larger voltages across it (current will suddenly begin when the electrons first force a fractal path through the crystal), and a mixing of metallic and insulating particles (the mixture can carry current only when the percolation limit is reached). In all of these cases, and in many others that could be mentioned, the fractal nature of the problem near the critical limit manifests itself clearly in the detailed behavior of the system.

A particularly important example from chemistry concerns the property of 'gelation', or the manner in which a fluid can suddenly lose its flowing properties as a function of the increase in size and entanglement of the molecules making up the fluid. For fluids of this particular type, the molecules increase in size by chemically linking one to another in a manner referred to as 'polymerization'. The progressive nature of this process (usually as water is removed from the fluid) finally leads to the formation of the 'infinite molecule' which, due to the degree of randomness involved in the process, is again of fractal form. The sudden change in fluid behavior upon formation of the infinite molecule (gelation) is reflected in all of the material properties, and a precise understanding of the process once again necessarily involves an appreciation of the fractal nature of the event. For the moment, regardless of the degree to which its details may be less than fully characterized, the technique is a valuable alternative means for forming glasses and other amorphous structures that cannot be prepared by the more conventional methods discussed in chapter 2.

Other important classes of fractals of interest to the scientific community are those which grow from a small core particle by repeated additions of identical particles that 'stick' at the point of random collision. The most familiar example to the non-specialist is probably soot (which consists of just such an aggregation of carbon particles), but others can be found in the electrolytic deposition of metals (electroplating), in colloid formation, and in corrosion. In each case, as the seed grows, tiny bumps or hollows form on the surface owing to the randomness of the sticking process. Further on in the formation, since random collisions are more likely to occur near the peak of a bump than in a valley or hollow, the growth rapidly assumes a less than solid-like appearance which, under the right conditions, can be a fractal with a fractal dimension typically of order $d = 2.5$. The advantage of this kind of fractal, to the theorist at least, is the comparative ease with which its formation can be modeled by the method known as diffusion limited aggregation, or DLA. This is particularly important because the dynamics associated with fractal formation in general are far less well understood than are the static geometrical aspects of the fractal patterns themselves.

The idea of the DLA method is to have particles diffusing towards a seed cluster by what is known as a 'random walk' process, and then sticking at the point of contact. The random walk aspect of the model arises because very small particles, when suspended in fluids, do not travel in naively anticipated Newtonian straight lines, but rather in a very haphazard and random-like manner. The earliest qualitative explanation of this phenomenon, which had been experimentally noted by a number of prior observers, was made by the Scottish botanist Robert Brown in 1827 following a study of microscopic pollen grains, powdered coal, and an assortment of other particles—all suspended in liquids. He deduced that it resulted from the varying numbers of fluid atoms bombarding an individual pollen grain as a function of time. For large particles the numbers involved are so large that the fractional variation of atoms pushing this-way-and-that is not sufficient to produce observable effects. However, for extremely small particles, these same variations can produce a most erratic motion which, seen under the microscope, consists of 'steps' in random directions with step lengths centered about some characteristic value. The quantitative theory was given by Einstein in 1906 and constituted the most direct proof, up to that time, of the existence of atoms. Today, the term 'random walk' is often used to describe this 'Brownian' motion. In the DLA computer simulation method, 'particles' are released one at a time onto a square grid from an area outside and allowed to diffuse by a random walk (with each step equal to one grid square) towards the origin where the seed is placed. Upon contact with the seed (or the ensuing cluster at any stage of its growth) the particle sticks at the position of contact. A typical aggregate of the kind formed by this sort of procedure is shown in figure 62—a fractal with a Hausdorff dimension close to $\frac{5}{3}$.

Objects of this kind seem to form in nature as a result of so-called 'non-equilibrium' growth. A perfect crystal, for instance, grows near equilibrium; it 'tries' many different configurations in an effort to locate the one with the most stable (i.e. lowest-energy)

Figure 62

structure. Thus, an equilibrium crystal grows slowly and is subject to constant rearrangement at the growth surface. DLA processes, on the other hand, do not have the luxury of time, and it is this fact (modeled by the instant sticking aspect of the growth process) that leads to the fractal nature of the result.

Even the dynamics of the random walk process itself can give rise to a fractal. An example in one dimension is easiest to envisage. Consider the situation in which a particle moves on a line starting from the origin at, say, $x = 0$. Suppose that it progresses one unit to the left or to the right with equal probability at each step. What will the resulting path look like? A little thought will convince you that this is just a physical manifestation of the coin-tossing problem of chapter 9, in which heads or tails gives way to steps to the left or right. From our earlier coin-tossing study we know that, after a very large number of steps n, the probability of reaching a point a distance k (steps) from the origin is proportional to the Gaussian function $\exp(-2k^2/n)$. The actual path is one of extreme jaggedness, and can be expected to pass back and forth through the origin many times. But what of its fractal dimension? Well, the Gaussian form tells us that if k

increases by a factor of two then n has to increase by a factor of four. Thus, measuring the path length P (i.e. n) in units of step length L we find

$$P(L/2) = 2^2 P(L)$$

indicative of a Hausdorff dimension $d = 2$. Here, therefore, we have a 'curve' which is topologically of dimension one and fractally of dimension two.

Examples of this phenomenon in nature usually consist of records of observations over time. Examples might be records of temperature variations, rainfall or the thickness of tree rings. These all exhibit erratic behavior on both short and long time scales and their traces are all (at least idealistically) fractals with many features in common with the random walk and with Brownian motion. The nuances of their deviations from true Gaussian behavior is still an area of active research in the field.

Even the Cantor set, with fractal dimension less than one, has been found to have common manifestations in physics. The most familiar is perhaps the 'noise' disturbance during radio broadcasting, an annoyance which can be persuasively modeled by Cantor sets. But the example I like best is that of a pendulum subjected to a stimulating force, such as an adult pushing a child on a swing. It is common knowledge (particularly to the child) that the motion takes on its simplest (and most enjoyable) form if the push, or stimulus, is in one-to-one phase with the frequency of the swing, or response. Less well known is the fact that, if the frequency of the stimulus is gradually (by which we imply continuously) changed, the frequency of the pendulum (or swing) will maintain its value for a while until, suddenly, the motion will become chaotic before settling down to a new frequency. This procedure then continues in a sequence of 'jumps' as the frequency of the stimulus continuously increases further. Thus, if we plot the frequency of the pendulum against the frequency of the stimulus, the resulting graph looks like a staircase.

The phenomenon is known as 'mode locking' and occurs in many different contexts in physics. The connection with fractals arises from the fact that the more carefully the experiment is performed, the more steps there are on the staircase. In fact, between any two steps there appears to be an infinitude of smaller steps, a phenomenon now referred to as the Devil's staircase. If the self-similarity of the structure is examined in detail under magnification, then the most striking feature is that it seems to possess a fractal dimension of about $d = 0.87$, no matter what its physical manifestation. The connection with the Cantor set is best appreciated by picturing the staircase and removing from the horizontal axis all except the sequence of points that mark the vertical 'jumps'. It is in the infinite limit that the resulting fractal 'dust' of points becomes a Cantor set.

The final indication that fractals have now truly become a fundamental part of solid state physics is the fact that, in the quantum theory of solids, their quantum units of atomic vibrational motion have been given their own name (fractons), and the theory giving rise to them has reached text book level. For very long wavelengths, atomic vibrational excitations, even in disordered materials, can be described in terms of 'acoustic phonons' or conventional sound-wave modes (see chapter 4). This arises because at long enough wavelengths all solids appear to be homogeneous or continuously smooth. However, at some shorter wavelength, the fractal nature of a fractal lattice will begin to become evident, and is conveniently observed in experiments (such as light scattering) that measure the frequency versus wavelength (or 'dispersion') relationship for the quantized vibrational modes. It is found that the mode dispersion and the general character of the scattered radiation change from their well known 'phonon' form to patterns that are dramatically different. The new quantized excitations are called 'fractons' and the cross-over length scale separating the phonon and fracton regimes is very clearly defined. Using these measurements of the dynamic properties of fractal lattices, it is possible to define additional fractal dimensions (different from, though often related to, the Hausdorff dimension) that characterize other aspects of the fractal nature. More and more it is becoming apparent that the Hausdorff dimension alone is insufficient to define all the invariant characteristics of fractals that experiment can probe. On the other hand, detailed theories for many of these new experimental characterizations are not yet well developed, and the question of how many *independent* 'fractal dimensions' is sufficient for their complete interpretation remains a question for debate.

12

Motion: from Zeno to Schrödinger

The idea that space is 'continuous' is one that is appealing to intuition but difficult to express in a quantitative fashion. Problems associated with this notion of continuity first began to cause embarrassment to mathematicians as long ago as the time of Pythagoras (sixth century BC). It was at this time that the mathematicians of Ancient Greece, in their attempt to explore the ratios and proportionalities of geometric lines, encountered an unexpected and rather major roadblock. It was the discovery that the ratios of the lengths of such lines (which they intuitively conceptualized as continuous) could not always be expressed within their number system, which contained all possible integers and their ratios (i.e. fractions). It was here, then, that they first ran up against the troubling concept of a number which cannot be expressed as a ratio of two whole numbers (see chapter 5). Specifically, they were able to demonstrate that the ratio of the hypotenuse of a right-angled isosceles triangle to the length of either of its (equal) shorter sides could not be expressed in this fashion.

In modern-day language we should simply say that the number in question, which we know and write as $\sqrt{2}$, is *irrational* and can be expressed (at least in principle) in decimal notation by a set of integers of infinite length that never repeats or cycles its digit pattern. For us, the only annoying fact is that we cannot write it down in practice and it is for this reason that we resort to a symbol like $\sqrt{2}$ to avert this frustration. To Pythagoras, however, for whom the integers and their ratios (that is, the rational number system) had previously been sufficient for the quantitative interpretation of *all* known phenomena, this was a major crisis since it implied that there were infinitesimally small 'gaps' of some sort between the numbers expressible as fractions. In particular, as regards $\sqrt{2}$, this implied that one can approach this number (on a 'number line') as close or closer than anyone might request by using rational fractions alone, but never actually reach it. A small difference between $\sqrt{2}$ and any neighboring rational number point must therefore always exist. In the 'limit' one can approach arbitrarily close, leaving only an 'infinitesimal gap' (whatever that might mean) but never actually get there.

Clearly the notion of continuity went beyond the rational number system in its complexity. But how was one then to picture the infinitesimal; this distance, for example, between $\sqrt{2}$ and its nearest rational number point? If an infinitesimally short length has any length at all, then an infinite number of them laid end-to-end should constitute a line of infinite length; on the other hand, if it has no length, then even an infinite number end-to-end will likewise have no length. How was it possible to conceive of a line of finite length being composed of infinitesimal segments?

The problem was brought into particular focus in the late-fourth and early-third centuries BC by the Phoenician philosopher Zeno. Though neither a mathematician nor a physicist, Zeno concerned himself with the problem of describing motion. The Pythagoreans had assumed that space and time can be described, respectively, in terms of points and instants, and had concerned themselves primarily with form and structure rather than with flux and variability. The most often quoted of Zeno's four paradoxes involves the flight of an arrow. An arrow flying through the air, he said, must at any instant be at one place (since it cannot be in two places at once). While it is in this place it is at rest there and cannot therefore be moving. How then can it ever fly through the air?

The solution, in principle, must lie in the development of a mathematical system capable of describing 'moving' pictures of the arrow, and not merely a series of unmoving pictures 'projected' in rapid succession. But such an idea immediately raises the problematic question of dividing by zero. For example, if we wish to define velocity at a point, and to approach the problem by representing it as the ratio of a small distance traveled (Δx) to the small time interval taken (Δt) as the arrow moves away from this point, then the velocity *at* the point in question requires a formulation of the ratio $\Delta x/\Delta t$ as both the numerator and the denominator tends to zero. But dividing anything by zero (even zero by zero) is not allowed within the framework of mathematics because it leads to inconsistencies. Zeno's paradox is perhaps best described in modern terms by a 'movie' taken of the arrow in flight. If the projector is stopped at any instant, we can stop the arrow in mid-flight to see exactly where it is at that instant. But once we do this, the picture frame is still and the motion has disappeared. Evidently, it is just as difficult to conceive of time as composed of an infinite succession of infinitesimal 'instants' as it was to assemble a finite line from an infinite number of infinitesimal segments. The infinitesimal, which is intimately involved in both these dilemmas, can neither have a zero magnitude nor a non-zero one. What then is it? The Greeks had no answer and promptly left the problem to mathematical generations yet to come.

It seems clear that the answer to the paradox of Zeno's arrow must involve notions of continuity and limits; abstractions related to number systems of a kind that the Greeks were never fully able to master. And there the problem of numerically defining motion remained for almost two thousand years, until the sixteenth century. The problem was then revived primarily by physicists who wished to measure velocities at a point with experimental precision. In doing so, however, they largely avoided the mathematical difficulties of principle by simply measuring a 'localized' average velocity in terms of the time taken to move between two closely spaced points which bracketed the point in question. The French philosopher and mathematician René Descartes (1596–1650) did go a little further in recognizing that for bodies moving with a constant velocity one could plot a curve of their motion (with distance on one axis and time on the other) such that the resulting depiction was a straight line, for which the *slope* was a measure of the velocity. But the more general picture for bodies moving with variable velocity required the mathematical description of the slope of a curve *at* a point — and raised once more the notion of the infinitesimal.

A few results for mean velocities in special cases were also known. As far back as the thirteenth century, a group of scholars at Oxford University had recognized that a freely falling body was uniformly accelerating. They were also able to describe this motion geometrically and demonstrated that the mean velocity was just the average of the initial and final velocities. Also, of course, Galileo Galilei (1564–1642), in conducting his renowned experimentation concerning the laws of motion, was greatly concerned with experimental determinations of both velocity and acceleration. But the mathematics necessary for confronting the concept of instantaneous velocity or acceleration was just not

Gottfried Wilhelm von Leibniz, 1646–1716. Reproduced by permission of Mary Evans Picture Library.

available to him. This was, however, soon to appear and was formulated, finally, in the second half of the seventeenth century in the form of 'the calculus'. The accomplishment is generally attributed to two men—von Leibniz and Newton—although few breakthroughs of this magnitude are solely the work of single individuals. The invention of the calculus (or the theory of fluxions, as it is also known) followed a long and somewhat uneven flow of philosophical struggles and mathematical demonstrations that included, over the years, many efforts almost identical to those finally achieved in the calculus itself. Nevertheless, the traditional honor goes to Newton and von Leibniz (jointly, although, at the time, a bitter controversy ensued as to the priority and independence of their respective efforts).

Let us demonstrate the treatment of the motion of the notorious Newtonian falling apple as an introduction to the theory of fluxions. Galileo, in his role as perhaps the first modern physical scientist, had already determined that if an object (apple?) fell

vertically from rest under the influence of gravity, then after a
time t (seconds) it would fall a distance $x = at^2$ (feet), where 'a' is a
constant (related to the gravitational acceleration at the Earth's
surface) approximately equal to 16 feet per second per second (or
16 ft s^{-2} as it is more commonly written). Taking this equation

$$x = -at^2$$

with the negative sign symbolizing the downward direction,
Newton envisaged the mathematical situation both at the time t,
and at a small time Δt later, by which time x had advanced to $x +
\Delta x$, where Δx represents a small distance. Consequently he could
also write, for time $t + \Delta t$,

$$x + \Delta x = -a(t + \Delta t)^2.$$

Expanding the square in this equation, and subtracting the
former equation for time t from it, now provides us with an
equation relating Δx to Δt as follows:

$$\Delta x = -2at(\Delta t) - a(\Delta t)^2.$$

Dividing both sides by Δt now leads to an expression for the mean
velocity in the time interval between t and $t + \Delta t$. It is

$$\Delta x/\Delta t = -2at - a(\Delta t).$$

But how short can we make the small time interval Δt? We can
certainly make it short enough that the quantity $a(\Delta t)$ on the
right-hand side of the last equation becomes extremely small
compared with $-2at$. In fact, the shorter and shorter we make it,
the closer and closer the mean velocity approaches the perfectly
well defined finite number $-2at$. On the other hand, we can never
make Δt exactly equal to zero, for then we are returned to the old
Zeno paradox. Newton proposed that both Δx and Δt be made
'infinitesimally small' (whatever that might mean conceptually)
in the sense that in this 'limit', the ratio $\Delta x/\Delta t$ (which is then
written as dx/dt) becomes *exactly* equal to $-2at$. This then is to
represent the velocity of the apple *at* the point x and time instant t
(with the negative sign implying that the velocity is in a downward direction).

From the start, these ideas proved to be immensely useful and,
although they did not really confront the logical problem of
Zeno's paradox, they did seem to give correct answers. But, in
spite of this success, the logical questions associated with the
conceptualization of the infinitesimals themselves (as opposed to
their ratio) still concerned Newton—so much so that, although
his original work on the calculus was completed in 1666, the first

published version only appeared (and even then almost incidentally) some ten years later. Newton placed all emphasis on the ratios, setting aside any logical objections to the meaning of the individual infinitesimals dx and dt themselves. In his famous *Principia Mathematica*, published in 1687, he writes 'ultimate ratios in which quantities vanish are not, strictly speaking, ratios of ultimate quantities, but limits to which the ratios of these quantities decreasing without limit approach . . .'. The focus, therefore, is placed fully on the ratios, with the true nature of the infinitesimals themselves (or 'ghosts of departed quantities' as they were mockingly referred to by critics) left to the philosophers.

The velocity $v = dx/dt$ is called the 'derivative' of x with respect to t, and the symbol dx/dt is actually due to Leibniz. Newton preferred the symbol $\dot{x}$, which is still in use, particularly when discussing derivatives with respect to time. Armed with the equation

$$v = dx/dt = -2at$$

it is now easy to repeat the procedure for finding a derivative to obtain

$$dv/dt = d(dx/dt)/dt = -2a.$$

This second differentiation of x with respect to t is usually represented by the symbol d^2x/dt^2, so that we finally arrive at the equation

$$d^2x/dt^2 = -2a$$

which is called the 'equation of motion' for the falling apple. Since $d^2x/dt^2 = dv/dt$ is a rate of change of velocity (that is, an acceleration) and $2a$ is just a constant, the above equation is perhaps the simplest of its kind, representing a motion under constant acceleration—in this case, that due to gravity.

What Newton and Leibniz achieved in progressing from $x = -at^2$ through $dx/dt = -2at$ to $d^2x/dt^2 = -2a$ is now referred to as an example of differential calculus. The opposite procedure, which in general (although not for this particular case) is more difficult to carry out, is called integral calculus and is immensely valuable in physics because most of the basic laws of motion are most easily expressed in terms of acceleration. The reason follows from the intimate relationship that exists between acceleration and force as set out by Newton in his laws of classical mechanics. They are three in number as follows:

1. If there are no forces acting on a body it will persist in its motion; that is, it will move in a straight line with constant velocity.

2. If there are forces on the body, then the rate of change of linear momentum (which is defined as mass times velocity) of the body will be equal to the force acting on it.

3. When two bodies act upon one another, the force due to the first upon the second is equal to, but in the opposite direction to, the force due to the second body upon the first.

If we designate the body to have a mass m, a position x, and a velocity v, then we can express Newton's laws in mathematical form. In particular, the first two become:

1. If force $F = 0$, $v = $ constant.
2. If $F \neq 0$, $F = m(dv/dt) = m(d^2x/dt^2)$.

If we compare this second law with the equation of motion for the apple we see now that the force on the apple must be $F = -2ma$. This is more commonly written in the gravitational context as $F = -mg$, with $g = 2a$ being the acceleration due to gravity.

The third law enabled Newton to prove certain conservation laws (for example, the conservation of linear momentum) when bodies interact, but we need not get into further details here. Suffice it to say that, from the second law, any motion for which we know (from experiment or theory) the relationship between force F and position x (say $F = f(x)$) must be governed by the equation of motion

$$m(d^2x/dt^2) = f(x).$$

Using integral calculus one can now derive the velocity $v = dx/dt$ and the position x explicitly as functions of time t, and the velocity at any particular point x follows without more ado provided only that the Newton–Leibniz method is valid.

To see how this works in practice, we shall spend a few moments 'solving' this problem for the case of the harmonic oscillator, which we have met before in chapters 4 and 7. This motion (an example of which might be a weight bouncing up and down on the end of a spring) has a force F proportional to the distance x from the static equilibrium position (or origin) $x = 0$ and of such a sign as always to tend to return the mass towards this origin. In short $F = -Kx$, where K is a positive constant. The equation of motion is therefore

$$m(d^2x/dt^2) = -Kx$$

where m is the mass involved. Those of you who know a little calculus will now quickly be able to integrate this equation for the 'motion' to obtain a solution (see appendix 2)

$$x = A \sin (nt)$$

in which $n = \sqrt{K/m}$ is the frequency and A is the amplitude of the oscillation. It follows that the velocity at time t is

$$v = dx/dt = nA \cos (nt)$$

and, since $\sin^2 (nt) + \cos^2 (nt) = 1$, the velocity *at the point x* is

$$v = n\sqrt{(A^2 - x^2)}.$$

Thus, not surprisingly, the motion has its greatest velocity when the mass passes through the origin ($v = nA$ at $x = 0$) and comes to rest at its extremes of amplitude ($v = 0$ when $x = A$ or $x = -A$).

A particularly important quantity possessed by any body of mass m and velocity v is its *kinetic energy T*, defined as equal to $\frac{1}{2}mv^2$ or, equivalently in terms of momentum $p = mv$, as

$$T = p^2/2m.$$

This is (as can be proved from Newton's laws) a measure of the work done on the body by the forces involved in the motion. In particular, for the harmonic oscillator (using the above expression for v) the kinetic energy can be written as

$$T = \frac{1}{2}mn^2(A^2 - x^2)$$

or equivalently (since $n^2 = K/m$, where K is the constant appearing in the equation of motion)

$$T = \frac{1}{2}K(A^2 - x^2).$$

Kinetic energy, as its name implies, is the energy of motion. We see from the above equation that for the oscillator it is maximum at the origin $x = 0$ and tends to zero at extreme extensions $x = A$ and $x = -A$ of the motion. Since there is no way that energy can be lost, or dissipated, from this system, the *total energy* (which we label E) must contain some additional contribution to compensate for these changes in kinetic energy. This additional energy is called the *potential energy* (which we give the symbol U) so that total energy conservation requires

$$T + U = E.$$

Recasting the kinetic energy expression for the harmonic oscillator in this form, we see that for such a system

$$U = \frac{1}{2}Kx^2$$

$$E = \frac{1}{2}KA^2.$$

Thus, at the origin $x = 0$ of the motion all the energy is kinetic, while at the extremes of amplitude all the energy is potential. We now note that $dU/dx = Kx$ is the negative of the force responsible for the motion $F = -Kx$. The implied relationship

$$F = -dU/dx$$

is not valid just for an oscillator, but can be established from Newton's laws as being completely general. Clearly, therefore, the Newtonian equation for motion $F = m(d^2x/dt^2)$ could also be written in the form

$$m(d^2x/dt^2) = -dU/dx.$$

It follows that an equation of motion can be written equally well by expressing the potential energy as a function of x as by characterizing the force.

By extending these kinds of equations to other contexts, and to three dimensions, all sorts of problems of motion could now be attacked by the Newtonian method. Most famous, at the time, were those concerning the forces of gravitation between the planets, which are known to depend on the inverse-square of the distance between each pair of planets. Although the many-body planetary set of differential equations of motion has provided a headache to this day (see chapter 7) the simplest two-planet version (i.e. Earth and Moon, or planet and Sun) quickly received exact solution. Newton himself provided the answer, showing that the path of each 'planet' is an ellipse (in agreement with astronomical observation to a high degree of accuracy).

From these beginnings, an enormous field of 'Newtonian' or 'classical' mechanics was quickly built up by applying Newton's equations to any problem involving forces and motion. Through all this, however, the foundations of the calculus were still under challenge; the problem of the meaning of the infinitesimals just would not go away. Leibniz and Newton tended to refute the criticisms by focusing on the often astounding agreement of the theory's results with experiment and they, in turn, often attacked the critics for their obsession with rigor (a strange stand for accomplished mathematicians). All attempts to provide a truly rigorous underpinning for the calculus failed until well into the nineteenth century, including those of perhaps two of the greatest eighteenth century mathematicians, Euler and Lagrange.

The first seemingly satisfactory definition of infinitesimals was given (in terms of the limiting behavior of continuous functions) by the French mathematician Augustin-Louis Cauchy. In particular, his definitions appeared to secure the logical foundations that were sought and they were largely accepted by his contemporaries. However, some looseness in these definitions was eventually pointed out and a more precise exposition (too complex for us to discuss in detail) was eventually formulated, with a large measure of the credit for this accomplishment going to the German mathematician Karl Weierstrass (1815–1897). The long task of providing a rigorous foundation for the calculus (completed, to the satisfaction of most, in the 1890s, but still a subject of active research into the 1960s) therefore dragged on for well over two hundred years after Newton and Leibniz first formulated the method. During this time most scientists and many mathematicians were happy to accept the calculus on trust. In fact, to this day virtually no scientists (even those who happily use the calculus on a daily basis) make any attempt to read, let alone comprehend, the rigorous underpinnings of the method.

None of these mathematical wranglings seriously interrupted, or even slowed down, the progress of Newtonian mechanics throughout the eighteenth and nineteenth centuries. Prominent among the workers in the field was a group of Swiss mathematicians led by Leonhard Euler and various members of the illustrious Bernoulli family. It was Euler who first put Newton's second law into the form in which it is now most commonly remembered—namely, force equals mass times acceleration. Together with the Bernoullis he showed how a wide range of mechanical problems could be formulated and solved using integral and differential calculus. Building upon the work of the Frenchman D'Alembert concerning the equation of motion for a vibrating string (see chapter 4), they worked on other examples of motion in continuous systems and, along the way, gave birth to the science of hydrodynamics (or wave motions in liquids).

Very soon, a unified and systematic approach to all problems in physics that involved motion was established. A system of formal equations was derived applicable for all types of multi-particle motions that were expressible in terms of the derivatives of the function $L = T - U$ (where T is the total kinetic energy and U is the total potential energy) with respect to position and velocity. The function L itself, which played the fundamental role in these equations, was known as the 'Lagrangian' after Joseph Louis Lagrange. The major difficulty with this formalism was the fact that not only were the equations of second order in the derivatives, but the principal variables (position x and velocity $v = dx/dt$ for each particle) were not independent quantities. The situation improved considerably when it became apparent that these same equations could be expressed as first-order differential equations in the function $H = T + U$ (called the 'Hamiltonian' of the system, after the Irish mathematician William Rowan Hamilton). This could be done if each second-order equation in L was replaced by two first-order equations in H. The Hamiltonian is therefore the total energy of the system, and the equations of motion in terms of it are important enough to our story to be given in full:

$$dx_i/dt = \partial H/\partial p_i$$
$$dp_i/dt = - \partial H/\partial x_i$$

where i refers to the ith particle (which is at position x_i with momentum p_i). An additional advantage of this formalism is the fact that x_i and p_i in the equations can now be treated as independent quantities since they are related via the equations themselves (a benefit gained by replacing every Lagrangian equation by two Hamiltonian ones). The symbol ∂ in the above is called a partial derivative and is carried out holding constant all variables except the one in the derivative itself, see appendix 2.

Although all of this, to the uninitiated, may seem to be unduly complicated, the underlying simplicity is immediately evident if we just apply it to the now familiar (single-particle) harmonic oscillator problem for which we have

$$H = (p^2/2m) + \tfrac{1}{2}Kx^2.$$

Hamilton's equations of motion now follow directly as

$$dx/dt = p/m$$
$$dp/dt = -Kx$$

from which it follows further that

$$m(d^2x/dt^2) = (dp/dt) = -Kx$$

which is our old faithful Newtonian equation of motion. Since the total energy function H for a system of particles in motion is usually very simple to formulate (just a $p^2/2m$ for each particle gives the kinetic part T, and the potential part U follows immediately once the forces involved are defined) the Hamiltonian equations are ideally suited for the classical study of motion.

The use of equations of this kind for the description of many-particle motion in physics led rapidly to successful descriptions of such dynamic properties of materials as sound transmission, thermal conductivity, viscosity and diffusion. Coupled with the probability arguments leading to statistical mechanics (see chapter 9) a complete bridge between the microscopic world of moving atoms and molecules, and the macroscopic world of bulk measurements and thermodynamics, was firmly in place by the end of the nineteenth century. The notion of velocity at a point seemed to be on a firm foundation, and an understanding of all physical properties starting from classical equations of motion at the atomic level appeared to be just a question of mathematics and perseverance. However, as the nineteenth century gave way to the twentieth, physicists suddenly found themselves forced to revise some of their most fundamental notions about the very nature of matter.

With the development of vacuum apparatus, radio techniques, and other technical aids, the end of the nineteenth century brought with it the discovery of electrons, x-rays and radio-activity, and the possibility of experimentally studying the properties of individual atomic particles. It then turned out that the classical physics derived from Newton's laws was quite unable to explain many of the phenomena 'seen' at the atomic level. The first indications that all was not well with classical theory came in connection with the interplay of light (or more generally electromagnetic radiation of all wavelengths) with electrons. For a century or more before, light had been success-fully understood (and represented mathematically) as a wave motion, the energy of which was a continuous variable that manifested itself as the intensity of the light. In classical wave theory, the absorption and emission of light are described by processes involving the acceleration or deceleration of electrons. The theory could therefore be used to predict the intensity versus wavelength dependence of the spectrum of radiation that would exist in equilibrium inside a cavity at a temperature T. The result was alarming, showing that the intensity of radiation in equilib-rium with such an enclosure would increase without bound in the high-frequency limit. Experimentally, the radiation energy went through a peak as a function of frequency and approached zero again at high frequencies. In 1900, the German physicist Max Planck (1858–1947) noted that, if both the vibrating electrons in the container (which emit and absorb the radiation) and the radiation itself are restricted to have only certain discrete energy values, rather than continuous ranges, then an agreement of theory with experiment could be obtained. The implication, therefore, was that light could be absorbed and emitted only in separate portions or 'quanta' which we nowadays call photons. Quantitative agreement with the experimental radiation curve was achieved if the energy E of the light quanta varied with the frequency n of the radiation (now more commonly symbolized v

when it, as here, assumes measurement in units of *cycles* per second) in the fashion

$$E = h\nu$$

where h, called Planck's constant, has the value of 6.626×10^{-27} g cm^2 s^{-1} or equivalently 6.626×10^{-34} W s^2.

There was an initial reticence by many to accept Planck's radical suggestion since it implied a complete revolution in physics, being incompatible both with Newtonian mechanics and with the electromagnetic (or wave) theory of light. In addition, interference and diffraction phenomena, widely studied in optics, indubitably demonstrated the wave character of electromagnetic radiation. However, before long, additional phenomena were also observed that appeared to be understandable only in terms of Planck's quanta. Foremost among these was Einstein's demonstration in 1905 that an explanation of the way in which electrons are knocked out of atoms in a metal surface by electromagnetic radiation was also in conflict with a wave theory of light but quantitatively in accord with Planck's equation for light quanta. As the years passed, both theory and experiment increasingly confirmed the fact that light possessed a strange dual nature, sometimes exhibiting the properties of waves and sometimes of particles. Einstein saw this wave–particle duality, combined with the fact that electrons in atoms appeared to be restricted to certain discrete energy levels, as concrete evidence that physics needed a new mathematical foundation.

The wave-particle duality changes our whole view of the nature of light. For example, experimental observations of the encounters of photons with matter established that, in spite of having no mass, they did have a well defined momentum given by

$$p = h/\lambda$$

where λ is the wavelength of the radiation (related to the frequency ν by $c = \nu\lambda$ where c is the velocity of light) and h is once again Planck's constant. It followed that the energy and momentum of light 'particles' were related to the frequency and wavelength of light 'waves' by this constant h, which soon was to become one of the most fundamental numbers in all of physics.

Because of their intimate association, it was clear that every important change in the theory of radiation would necessarily involve a corresponding change in the theory of the electronic structure of matter. It followed that a satisfactory 'quantum theory' would have to apply to both radiation and matter.

However, up to 1924 it was thought that the problem was chiefly one of energy relations. Radiation was admittedly difficult, because of the wave–particle duality, but at least the behavior of electrons in atoms was to be described by the ordinary dynamical theories of particles, even if their orbitals were to be restricted in some manner by quantum conditions. But even this crumb of comfort was swept away by the PhD thesis work of a French physicist of noble birth, Prince Louis de Broglie, in 1924. De Broglie, inspired by Einstein's work on the wave–particle duality of light, became convinced that this duality was not just a property of electromagnetic radiation, but that it should be a property of *all* matter on the atomic scale. He proposed that every material particle should have associated with it certain wave properties. In essence, he said, instead of taking a wavelength (for light) and using it to calculate the energy and momentum of an associated particle (the photon) via the equations

$$E = h\nu \qquad p = h/\lambda$$

why not take the energy and momentum of a particle (such as an electron) and use these same relationships to calculate the frequency and wavelength of an associated wave? De Broglie himself had no experimental evidence at hand to support the revolutionary notion but, fired by his suggestion, experimenters soon rushed off to carry out the necessary tests. By 1927 and 1928 the evidence was available; under the right circumstances electrons do indeed behave like waves. In particular, when passing through thin metallic foils they were diffracted exactly as if they possessed a wave nature in accord with the above equations. Later on, this same wave nature was confirmed for other heavier atomic particles like neutrons and protons, and finally even for whole atoms. It was clear that de Broglie was right; in the world of the very small, particles and waves are twin facets of all entities.

Thus, de Broglie's proposal of a sweeping symmetry for physics, with light behaving like particles as well as waves, and electrons behaving like waves as well as particles, was correct. But what could the equation of motion of such mixed entities be, and how would its solutions relate to the classical motion of Newton? The answer was provided surprisingly quickly after the de Broglie conjecture (and interestingly before its experimental confirmation), in the spring of 1926, by the Austrian theoretical physicist Erwin Schrödinger (1887–1961). The equation is known as the 'wave equation' and it is the foundation of what is now referred to as 'wave mechanics'.

We can't derive this equation formally. Indeed, there is no

Erwin Schrödinger, 1887–1961. Nobel Prize for Physics 1933.
Photograph by Francis Simon. Reproduced by permission of
AIP Emilio Segrè Visual Archives.

derivation, any more than there is a derivation of Newton's
equations of motion—but the reasoning (or clever guessing)
behind Schrödinger's Nobel prize winning work can be ascer-
tained. The idea was based on a mathematical similarity between
mechanics and geometrical optics (or the theory of light ray
propagation) noted some 90 years earlier by William Rowan
Hamilton. The theory of light rays is a special limiting case of the
theory of light waves, and Schrödinger reasoned 'Why should
there not be a theory of particle waves that bears the same
relationship to particle trajectories as light rays have to light
waves?'

Beginning with the motion of a free particle, he defined the
associated 'matter wave', whatever that might turn out to mean,
by a variable (or 'wave function')

$$\phi = \exp\{2\pi i[(x/\lambda) - (t/T)]\}$$

which, recalling that $\exp(iz) = \cos(z) + i\sin(z)$ where $i = \sqrt{-1}$, is
a sinusoidal wave traveling in the direction x with wavelength λ,

frequency $1/T$ and velocity λ/T. Using the de Broglie relationships $T = h/E$ and $\lambda = h/p$, this can be transformed to

$$\phi = \exp[(2 \pi i/h)(px - Et)]$$

in which E and p are the energy and momentum of the particle to be associated with the wave function. By partial differentiation with respect to x and t we now find

$$(h/2 \pi i)(\partial\phi/\partial x) = p\phi$$

$$-(h/2 \pi i)(\partial\phi/\partial t) = E\phi$$

so that, within this mathematical framework, momentum is equivalent to the derivative (or 'operator') $(h/2 \pi i)(\partial/\partial x)$ and energy to $-(h/2 \pi i)(\partial/\partial t)$. Using these associations, the energy equation $E = T + U = (p^2/2m) + U$ now takes on the form

$$(h^2/8 \pi^2 m)(\partial^2\phi/\partial x^2) - U\phi = (h/2 \pi i)(\partial\phi/\partial t)$$

which is referred to as the (time-dependent) wave equation.

The validity of this equation naturally rests entirely upon its ability to describe the observed behavior of the microscopic world. However, such has been its success that it is now believed to determine (apart from relativistic corrections when necessary) the behavior of all material systems, although its predictions only differ significantly from those of Newtonian physics in the realm of the very small, for which the wavelength of the 'matter wave' is comparable to the size of the particle it describes. Because of the extremely small value for the Planck constant h, even a dust particle traveling at typical dust particle speeds is wholly macroscopic, with a matter wavelength a billion times smaller than the linear dimension of the dust particle. At the other extreme, an electron (which is thought to be essentially a point particle) accelerated through an electric potential of a few hundred volts has a matter wavelength comparable to the atomic spacing in solids, and can therefore be used to probe crystal structures in much the same way that x-rays are used for the same purpose.

Our starting free-particle wave function as set out above is actually the solution of Schrödinger's wave equation only for the case $U = 0$. For other cases that involve forces on the particle, U is in general a function of x, and solutions must be found by hard work and, for all but the very simplest cases, by numerical computation. However, for all systems with a constant total energy E (called 'conservative' systems), it is easy to see that the solution must be of the form

$$\phi(x, t) = \phi(x) \exp(-2\pi i E t/h)$$

where $\phi(x)$ is the solution of the time-independent wave equation

$$(\partial^2\phi/\partial x^2) + (8\,\pi^2\,m/h^2)(E - U)\phi = 0$$

which is perhaps the most famous of all twentieth century equations. Most simply it can be written in the so-called Hamiltonian form

$$H\phi = E\phi$$

where

$$H = (p^2/2m) + U$$

and p is simply replaced by its quantum operator equivalent

$$p = (h/2\pi i)(\partial/\partial x).$$

A generalization to three dimensions is straightforward and it is in this form that Schrödinger made his first application—specifically for the motion of an electron about the nucleus of a hydrogen atom. The hydrogen atom is the simplest of all atoms, consisting of a single electron moving around a central (much more massive) nucleus. The electron possesses a potential energy $U = -e^2/r$, which results from the attractive electrostatic force between the charge $+e$ on the nucleus and its counterbalancing charge $-e$ on the electron, where r is the distance between them.

This problem was of fundamental significance since, as we have described, it had been known from the turn of the century that atoms absorbed or emitted light for only very special frequencies. Before Schrödinger, it had been necessary to postulate that for some reason the orbiting electron could only exist in a select number of very special orbits with particular energy values. Newtonian theory was quite unable to explain this situation. In classical mechanics, the resulting orbitals are just those analogous to planets orbiting the sun and they can readily be defined with any of a continuum of energy values. It was therefore of immense interest to see what the wave equation predicted for the same situation.

The mathematics was by no means trivial, but the relevant three-dimensional wave equation could be solved in terms of known mathematical functions, and the task was made easier by the fact that the solution for the most difficult part had already been worked out in another context and published a couple of years before in a book entitled *Methods of Physics* by Richard

Courant and David Hilbert. Schrödinger found that the wave function ϕ remained finite over all space (even at infinite values for r) except for particular values of total energy $E = E_1, E_2, E_3 \ldots$, etc. Arguing that physically the electron cannot stray infinitely far from the nucleus with finite probabilities (at least for the smaller values of total energy) he was led to retain as physically meaningful only these special values for E and their respective wave functions $\phi = \phi_1, \phi_2, \phi_3, \ldots$, referring to the energies E_i as 'eigenvalues' and their associated wave functions ϕ_i as 'eigenfunctions'. The triumph of the wave equation emerged when the eigenvalues E_i were found to correspond exactly to those necessary for explaining the origin of the observed sharp-line optical spectrum of hydrogen in terms of photons being absorbed or created when the electron hops from one 'eigenlevel' to another. The wave equation therefore suggests that the electron can persist only in very special 'orbits' in which it is completely stable until some interaction with an outside entity causes it to absorb or emit a photon.

Schrödinger had evidently found a very persuasive method of extracting the 'allowed' energy levels of a hydrogen atom from first principles, and the success was soon extended to the description of more complicated atoms and to other contexts as well. But in spite of these enormous successes, fundamental problems of interpretation remained. For example, in the case of the hydrogen atom, what was the 'orbit' of the electron in any of these eigenlevels? In his first papers, Schrödinger spoke of the wave function ϕ as a vibrational amplitude in space, and sought to interpret the real number quantity $\phi^*\phi$ (where $\phi^* = a - ib$ if $\phi = a + ib$ is complex) as the electric charge density. But since experiments clearly indicated that the electron was a strongly localized particle, while ϕ and its 'square' $\phi^*\phi$ are quite diffuse, there was clearly some interpretational confusion intermingled with the numerical successes. For example, exactly what is it that is vibrating in a matter wave? Since ϕ is, in general, a numerically complex quantity, it is inherently more difficult to visualize than are the real electric and magnetic fields associated with a light wave.

Shortly thereafter, the German-born (and later naturalized British) physicist Max Born (1882–1970) interpreted the square of the wave function as a probability, and renounced as meaningless in principle any attempt to go beyond probabilities to arrive at a theory involving the notion of localized electrons traveling in orbitals. Over the years this statistical interpretation has become the accepted dogma of quantum theory although a number of extremely distinguished scientists (Schrödinger and Einstein among them) were never happy at this retreat from determinism.

If we go back to the wave function for a free particle (traveling in direction x with momentum p), we note that ϕ is a complex sinusoidal function for which $\phi^*\phi$ is completely independent of x. It follows, according to the probability interpretation, that this particle can be found with equal probability *anywhere* along its straight line trajectory. An exact knowledge of momentum, and therefore velocity, has resulted in a complete absence of knowledge of position. If, however, we allow the particle to have a distribution of momenta about p, and represent the wave function by a sum of sinusoidal waves with slightly different momenta, then (along the lines of the Fourier summations discussed in chapter 4) we can produce a 'wave packet' that is localized to a degree about a particular spatial position x. However, we can only define a wave function truly localized at a point if we incorporate waves that utilize equally all possible values for momentum. Quantum mechanics seemingly deters us from obtaining simultaneously an exact knowledge of both particle position and momentum. Expressed in another way, we can never know exactly the velocity of a particle at a point; Zeno's paradox has therefore seemingly returned, in spite of all the efforts of Leibniz and Newton to rid us of it.

In 1927, the German physicist Werner Heisenberg (1901–1975) established the exact limits that quantum theory inescapably places on the accuracy with which momentum and position can be simultaneously determined. (Interestingly, Heisenberg had also solved the problem of discrete energy levels, in his case in the context of the harmonic oscillator, as early as the summer of 1925, and Wolfgang Pauli (1900–1958) had adapted it for the hydrogen atom a few months before Schrödinger. However, the method used was based on matrix theory and, although the two methods are now known to be equivalent, Schrödinger's was much more widely used in the early days since most physicists were far more familiar with differential equations than with matrix representations. In addition, Schrödinger's approach seemed to most to convey a more definite physical picture than did the matrix method.) In what is now known as Heisenberg's uncertainty relation, Heisenberg showed that if an attempt is made to measure both the position x and momentum p of any particle, then the product of the average errors of measurement, symbolized by Δx and Δp, is greater than or equal to $h/4\pi$, where h is Planck's constant. In symbols we can therefore write

$$(\Delta x)(\Delta p) \geq h/4\pi.$$

There are also several other important pairs of measurables that the very nature of quantum mechanics prevents us from determining simultaneously with arbitrary accuracy and for which analogous uncertainty relationships can be set out. However, we shall concentrate on that involving position and momentum since this is the one most relevant for the discussion of motion.

One assumes in classical physics that for any state at any time a particle has well defined values for x and p. In quantum mechanics the Heisenberg uncertainty relationship tells us that this is just not so. The momentum of a particle is no longer a function of the coordinates of the particle but is a property of the wave function as a whole, and the classical definition of momentum as $p = m(dx/dt)$ just cannot be applied to objects in the quantum world. The reason that these peculiar effects are not noticed in the macroscopic world is, of course, the small size of the Planck constant h. If, for example, we return to our dust particle (which is about as small as the objects of our everyday experience get) with a typical dust particle velocity (say a few mm s^{-1}), then we find that if its position is measured to one hundredth of its diameter, then the quantum uncertainty in momentum is about a billion times smaller than its classical momentum.

If you consider possible experiments that might be designed to measure x and p for a quantum particle like an electron, then a possible source of the difficulty spelled out by the inequality $(\Delta x)(\Delta p) \geq h/4\pi$ can be deduced. For example, one could measure the position x of such a particle by bouncing photons off it (in the same way we see normal objects). But to 'see' it clearly it is necessary to use radiation with a wavelength smaller than the particle and, for a quantum particle of any kind, this requires the use of very-short-wavelength photons. But short-wavelength photons, by the de Broglie relation, have large momenta and would therefore seriously disturb the particle they are designed to locate and make it impossible to determine p with arbitrary accuracy. One might therefore, perhaps, be tempted to interpret the uncertainty relation merely as a restriction on the observer's knowledge and not necessarily feel forced to abandon the notion that a definite position and momentum (or velocity) exist. But theories based on properties that cannot be observed are dangerous and all too often tend to lead to frustration and inconsistency (as witnessed by all the time and effort wasted on the hypothetical 'luminiferous ether' through which light was perceived to propagate in pre-relativity days). In recognition of this fact, the accepted interpretation today is that the physical world possesses only those properties that can be revealed by experiment. The implication is that theory can deal only with the results of observation and not with any 'underlying reality' that may or may not exist.

It is not appropriate here to pursue the enormous successes that Schrödinger's wave equation has led to in physics over the subsequent decades nor upon the philosophical questions that are still raised from time to time concerning the interpretation of the wave function itself. We shall merely demonstrate a few findings of the theory that would have been of particular interest to Zeno, and set them out in the context of our old friend the harmonic oscillator which was discussed earlier in its Newtonian guise. Substituting the relevant potential energy $U = \frac{1}{2}Kx^2$ into the wave equation, and solving subject to the condition that ϕ must tend to zero at an infinite distance from the origin $x = 0$, once again forces the total energy E (and wave function ϕ) to take on a set of discrete values (and functional forms). Specifically the energies are restricted to values

$$E_n = (h\omega/2\pi)(n + \tfrac{1}{2})$$

where $\omega = \sqrt{K/m}$ is the classical frequency (now in units of radians per second) of the oscillator, and the 'quantum number' n can take all integer values from $n = 0$ up to arbitrarily high values. The associated wave functions can be written in the form

$$\phi_n(\xi) = H_n \exp\left(-\tfrac{1}{2}\xi^2\right)$$

where $\xi = x\sqrt{(2\pi m\omega/h)}$, and H_n is a polynomial (called a Hermite polynomial after one of France's most influential nineteenth century mathematical analysts Charles Hermite (1822–1891)). The explicit form of these polynomials for the first few values of n are

$$H_0 = 1$$
$$H_1 = 2\xi$$
$$H_2 = 4\xi^2 - 2$$
$$H_3 = 8\xi^3 - 12\xi.$$

These wave functions ϕ_n and their 'squares' $\phi_n{}^*\phi_n$ (adjusted in amplitude so that the areas under the latter curves are always equal to unity in accord with Born's probability interpretation) are shown for the first few values of n in figures 63 and 64. Also

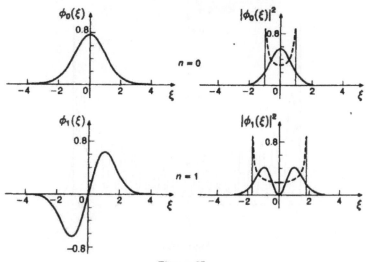

Figure 63

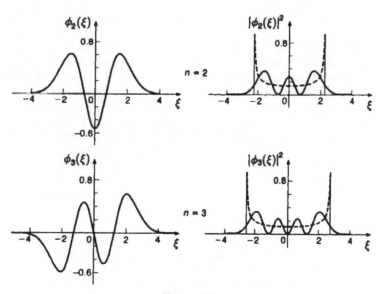

Figure 64

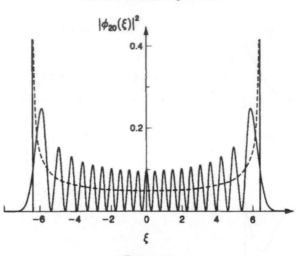

Figure 65

shown, as dashed curves in these figures, is the probability distribution for the classical Newtonian oscillator. We see that the probability that the oscillating mass m will be found at any particular value for ξ (or correspondingly for x) is very different in the lowest quantum states $n = 0, 1, 2, 3$ from what one would expect for the classical oscillator. We note, in particular, that there is even a finite probability of finding the oscillating particle at distances beyond the classical amplitude (an effect known as 'quantum tunneling'). However, it is anticipated that the quantum description will approach closer and closer to the classical probability distribution as n increases towards infinity because this limit, for a constant value of total energy E, is equivalent to letting Planck's constant tend to zero, and hence for quantum effects to disappear. In figure 65, we show the probability distribution for $n = 20$, from which you can begin to see how the approach to the (dashed) classical limit is going to take place.

Let us now look a little more closely at the lowest quantum energy level of all. Classically we should expect the lowest energy

$$E = p^2/2m + \tfrac{1}{2}Kx^2$$

to be $E = 0$, corresponding to the 'at rest' position with $x = p = 0$. But the uncertainty principle will not allow this state to exist quantum mechanically since it implies an exact simultaneous knowledge of position and velocity. It follows that the quantum oscillator can never be at rest. In fact, the mean values for x^2 and for p^2 (usually symbolized as $\langle x^2 \rangle$ and $\langle p^2 \rangle$ respectively) can be calculated for the ground state $n = 0$ by integration over the probability function $\phi_0^* \phi_0$, when we find

$$\langle x^2 \rangle = h\omega/4\pi K \qquad \langle p^2 \rangle = hm\omega/4\pi.$$

It follows that displacement and momentum contribute equal amounts to the ground-state energy $E_0 = h\omega/4\pi$ in the manner

$$\langle p^2 \rangle/2m = \tfrac{1}{2}K\langle x^2 \rangle = h\omega/8\pi.$$

We also now note that the implied quantum uncertainties of position Δx and momentum Δp from their classical 'ground-state' analogs $x = p = 0$ follow as

$$(\Delta x)(\Delta p) = \sqrt{\langle x^2 \rangle \langle p^2 \rangle} = h/4\pi$$

in accord with the equality in Heisenberg's uncertainty relationship.

It may therefore be comforting to know that the pendulum on your grandfather clock can never 'stop' in the true quantum mechanical sense even if you fail to wind it, although the associated energy (called the 'zero-point' energy) is unfortunately far too small to be of any consequence in the macroscopic world. Nevertheless, the presence of this zero-point energy is quite real and, in the world of the atom, its consequences can be experimentally observed in many different contexts. In fact, in some circumstances it is this energy that determines the stable symmetry state of certain crystal structures.

The situation therefore seems to be far worse than even Zeno imagined. Zeno was happy to accept the reality of an instant of time at which a moving oscillating particle could be 'frozen' at rest, as in a movie frame. He was baffled only by the difficulty of reconciling this picture with the notion of velocity at a point. In the world of Schrödinger we cannot even locate the particle exactly at any time, much less worry about its velocity. All is now wrapped up in the mysteries of wave–particle duality and Born's probability interpretation. The 'official position' is, as stated above, that the physical world has only those properties that can be revealed by experiment (including both the wave and particle aspects of matter) and that any attempt to go further than this is not only idle speculation, but is likely to lead to both confusion and inconsistency. A particle is an unobservable wave—a purely mathematical entity representing all of the possible states of the particle—until it is observed. At that point, and at that point only, does the wave somehow 'collapse' into a 'localized' particle.

Not everyone is happy with this seeming abandonment of physical reality, causality and determinism and its replacement with the language of chance. Perhaps the most famous quote from the disaffected was Einstein's 'God does not play dice'. But many others have retained at least a modicum of dissatisfaction with the *status quo* to the extent that there is still a considerable amount of activity in the whole area of trying to understand quantum theory better. Efforts to construct deeper deterministic levels of description, though fraught with danger, have been attempted. In these theories, quantum mechanics remains intact but is to be understood in terms of a more detailed picture. The theories are generally referred to as 'hidden-variable' theories since they postulate details that go beyond the observable, and a fair analogy is that a successful hidden-variable theory would be to quantum mechanics as classical statistical mechanics is to classical mechanics.

It is absolutely essential, of course, that the predictions of such theories for observables never disagree with quantum theory and this restriction places severe conditions upon the nature of hidden-variable theories. In fact, for many years an erroneous 'proof' by the German–American mathematician John von Neumann (1903–1957) that such a task could not be accomplished remained unchallenged, and severely discouraged efforts in the field of hidden variables. More recently, however, theorems by the British physicist John Stewart Bell (1928–1990) have spelled out the conditions that have to be met. They are severely restrictive, but some attempts at hidden-variable theories which are not in any violation of Bell's restrictions have already been made, and some believe that they provide a more satisfactory account of the physics than does the conventional statistical interpretation.

If all this seems confusing to you, then you are in good company for it was the Danish physicist Niels Bohr (1885–1962), whose own contributions to the development of quantum theory were more than sufficient to win him a Nobel prize, who said 'If you aren't confused by quantum physics, then you haven't really understood it'. And to make matters even worse, it seems that we can never safely assume that even the conventional statistical interpretation is necessarily 'set in stone'. Even as I write, new experiments are being planned which might, at last, enable a measurement to be made of the wave function itself. If this effort is successful it will necessitate a whole re-evaluation of what is measurable in the physical world. The quest for an understanding of velocity at a point may indeed not yet be over.

Appendices

Appendix 1

In order to 'translate' the geometry of three-dimensional space into algebra, it is first necessary to label any point in that space by numbers (or more generally letters). For three dimensions a point requires three letters (say x, y and z) since, after choosing an arbitrary reference point, or 'origin', for which $x = y = z = 0$, any other point P can be labeled by three numbers x, y, z measuring, in geographic terms, latitude, longitude, and elevation with respect to the origin. We therefore write a point P as P(x, y, z) and refer to the measures x, y, and z respectively as the 'coordinates of P along the x-, y- and z-axes'.

If we now consider a second point Q($x + a$, $y + b$, $z + c$) in the immediate vicinity of P then, if the space is Euclidean, we can relate Q to P geometrically by the construction shown in figure 66. The construction involves a box-shaped volume (called a rectangular parallelepiped in the language of solid geometry) with 90 degree angles, and side lengths a, b, and c as drawn. Making a double use of the theorem of Pythagoras, as detailed in the figure, it is clear that the distance L between P and Q (that is, the body diagonal of the rectangular volume) can be expressed as

$$L^2 = a^2 + b^2 + c^2.$$

If a, b, and c are now allowed to become infinitesimally small quantities (in a sense defined in more detail later in the book) it has become customary to label them in a special manner to denote this fact; specifically as dx, dy, and dz, and to label the now infinitesimally small distance between P and Q as ds. No multiplication between d and x (or y, z, or s) is implied by this 'd-

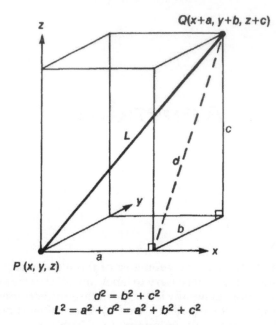

$$d^2 = b^2 + c^2$$
$$L^2 = a^2 + d^2 = a^2 + b^2 + c^2$$

Figure 66

symbolism'—it is simply a two-letter notation to represent an infinitesimally small distance in the relevant direction. Thus, in this so-called 'differential' formalism, the distance ds between what are now referred to as 'neighboring' points P and Q in Euclidean space is

$$ds^2 = dx^2 + dy^2 + dz^2$$

as given in the text on page 45. A more thorough discussion of precisely what is meant by an infinitesimal is given in chapter 12.

Appendix 2

Although nowhere in this book shall we require the reader to confront methods of solution of differential equations, it is perhaps helpful to those unfamiliar with differential symbolism for us to set out here a short introduction to the concepts and vocabulary involved. This will be done with the aid of simple examples and, although the treatment is obviously neither rigorous nor complete, it should assist the non-expert to a better understanding of the text and enable him or her, at least, to verify the solutions given therein.

If f is a function of a single variable (say x) as, for example, in

$$f = x^3$$

then changing x by an infinitesimal amount dx leads to a corresponding change df in f given by

$$f + df = (x + dx)^3.$$

Expanding the cube (using the binomial theorem) and subtracting the original equation from the result leads to

$$df = 3x^2 dx$$

where we have omitted terms in $(dx)^2$ and $(dx)^3$ since they are in general negligibly small compared with dx. Dividing both sides by the infinitesimal dx now gives

$$df/dx = 3x^2.$$

We say that we have 'differentiated' the function f with respect to x and obtained the 'first derivative' df/dx of f. In geometric terms df/dx measures the slope of f at the point x, and the equation $df/dx = 3x^2$ is an example of a 'first-order differential equation', one solution of which, from the above derivation, must be $f = x^3$.

If we now define $g = df/dx = 3x^2$ we can proceed in the same fashion to calculate $dg/dx = 6x$. The quantity dg/dx is, by its definition, $d(df/dx)/dx$. This is conventionally denoted by the symbol d^2f/dx^2, called the 'second derivative of f with respect to x, and geometrically measures the curvature of f at the point x. We may therefore write, for our example,

$$d^2f/dx^2 = 6x$$

which is a 'differential equation of second order'. Once again, we know that at least one of its solutions must be $f = x^3$.

If f is a function of more than one variable, say $f = x^3 t^2$, then by letting x and t change, respectively, by infinitesimal amounts dx and dt we obtain

$$f + df = (x + dx)^3 (t + dt)^2.$$

Expanding the right-hand side (neglecting terms that multiply two or more infinitesimals together) and subtracting the original equation now leads to

$$df = 3x^2 t^2 dx + 2x^3 t \, dt.$$

In differential formalism this is conventionally written as

$$df = (\partial f/\partial x)dx + (\partial f/\partial t)dt$$

where $\partial f/\partial x$ and $\partial f/\partial t$ are called the 'partial derivatives' of f with respect to x and t, respectively. Thus, we have

$$\partial f/\partial x = 3x^2t^2 \qquad \partial f/\partial t = 2x^3t$$

leading to an obvious relationship

$$2x(\partial f/\partial x) = 3t(\partial f/\partial t)$$

since both sides are equal to $6x^3t^2$. The above equation is an example of a first-order 'partial' differential equation. Geometrically, a function of two variables is a surface. Clearly the 'slope' of such a surface is not defined until you specify the direction along the surface that is of interest. For $f = x^3t^2$, the partial derivative $\partial f/\partial x$ measures the slope at the point (x, t) in a direction that keeps t constant. Analogously, $\partial f/\partial t$ measures the slope for constant x.

Clearly one can now go on to define higher-order partial derivatives in the forms

$$\partial^2 f/\partial x^2 = \partial(\partial f/\partial x)/\partial x = 6xt^2$$

$$\partial^2 f/\partial x\partial t = \partial(\partial f/\partial x)/\partial t = 6x^2t$$

$$\partial^2 f/\partial t^2 = \partial(\partial f/\partial t)/\partial t = 2x^3$$

from which one can write second-order partial differential equations like

$$x^2(\partial^2 f/\partial x^2) = 3t^2(\partial^2 f/\partial t^2).$$

Since we have specifically prepared this equation from $f = x^3t^2$, it follows that we already know at least one of its solutions. In physics, equations of this kind are usually derived directly from laws of motion, and their general solution can sometimes be difficult. However, it is usually simple to test any particular function to see whether it satisfies the equation. To do this it is only necessary to know what the derivatives of common functional forms are. They can always be worked out 'longhand' by methods paralleling those given above. In practice, however, they can usually be found tabulated in books on mathematical functions. In this book the most common derivatives required to verify the given solutions of the differential equations that appear are as follows:

Function $f(x)$	df/dx	d^2f/dx^2
x^k	kx^{k-1}	$k(k-1)x^{k-2}$
$\sin(kx)$	$k\cos(kx)$	$-k^2\sin(kx)$
$\cos(kx)$	$-k\sin(kx)$	$-k^2\cos(kx)$
e^{kx}	ke^{kx}	k^2e^{kx}

where k is a constant. Partial derivatives can be obtained as ordinary derivatives so long as all variables except the one involved in the differentiation are treated as constants.

Appendix 3

The task at hand is to demonstrate that the equations

$$a^3 - 3a + 1 = 0$$
$$b^3 - 3b + 1 = 0$$
$$c^3 - 3c + 1 = 0$$

remain satisfied when a, b, c are replaced respectively by $c^2 - 2$, $a^2 - 2$, and $b^2 - 2$. Making these replacements directly on the left-hand sides of the above equations leads to the functions

$$c^6 - 6c^4 + 9c^2 - 1$$
$$a^6 - 6a^4 + 9a^2 - 1$$
$$b^6 - 6b^4 + 9b^2 - 1$$

where we have made use, in turn, of the binomial expansion

$$(x + y)^3 = x^3 + 3x^2y + 3xy^2 + y^3$$

with $x = c^2, y = -2; x = a^2, y = -2; x = b^2, y = -2$. The resulting functions are each factorizable into two cubics in the form

$$(c^3 - 3c + 1)(c^3 - 3c - 1)$$
$$(a^3 - 3a + 1)(a^3 - 3a - 1)$$
$$(b^3 - 3b + 1)(b^3 - 3b - 1)$$

as may be verified by directly expanding these products. Each is therefore still equal to zero by virtue of its first factor, as can be seen from the original equations. It follows that the original equations remain valid under the said transformations.

Index

9780750301039

An environmentally friendly book printed and bound in England by www.printondemand-worldwide.com

PEFC Certified

This product is
from sustainably
managed forests
and controlled
sources

PEFC
PEFC/16-33-415

www.pefc.org

This book is made of chain-of-custody materials; FSC materials for the cover and PEFC materials for the text pages.

#0480 - 240216 - C0 - 234/156/16 - PB - 9780750301039